ROBOTS IN MANUFACTURING

Key to International Competitiveness

Jack Baranson, Ph.D.
President
Developing World
   Industry and Technology, Inc.

Lomond Publications, Inc.
Mt. Airy, Maryland 21771
1983

Library of Congress Catalog Number: 83-81240
ISBN: 0-912338-39-3 (clothbound)
        0-912338-40-7 (microfiche)

Printed in the United States of America

Published by:
Lomond Publications, Inc.
P.O. Box 88
Mt. Airy, Maryland 21771 U.S.A.

PUBLISHER'S FOREWORD

As Dr. Jack Baranson points out, robotics and robotization are at the threshold of very fundamental change in manufacturing, productivity, world and national economics, and industrial policy and management. National roles in world industrial markets depend to a significant degree on successful development and installation of automatic manufacturing equipment and systems (AMES).

No one is better qualified to address this matter than Dr. Baranson, and he does so in this book with lucidity, candor, insight and conviction. We believe publication of his research findings and analysis is a significant service to policy makers and to interested citizens.

Dr. Baranson's intimate, on-site knowledge of the industrial economies of the world through his work as economist with the World Bank and his private consultancy to government and business provides him with authoritative insights into the industrial and economic dynamics of robotization.

The nature, transfer, application, management and impact of technological change is a major theme of Lomond books and monthly serials. We are indeed pleased to be able to publish this timely, enlightening and very valuable message.

Lowell H. Hattery
Publisher

PREFACE AND ACKNOWLEDGEMENTS

The looming importance of robotics—and the broader field of automated manufacturing—is evidenced by the increased frequency of relevant full length media articles and special television coverage. There is a growing awareness that one of the root causes of the competitive gap that now exists between the United States and Japan in certain industrial sectors is the failure on the part of U.S. industry to invest in the new and highly productive Automated Manufacturing Equipment and Systems (AMES) as rapidly and extensively as have Japan and some segments of the European economy. The decline in international competitiveness of large segments in our economy has become a cause of deep concern to both businessmen and government officials.

U.S. government officials are especially concerned over the contraction of our industrial base, which has resulted in part from the inroads of foreign competitors, and given rise to problems of industrial unemployment, budget deficits, and balance of payment pressures. The current high level of industrial unemployment poses a difficult political and economic dilemma: the increase in the rate of introduction of AMES necessary to enhance our competitive position and provide job opportunities in the long run could well further eliminate industrial jobs in the short run.

Some of our largest firms in an ever-widening array of high-technology products—automotive and heavy mechanical equipment, consumer electronics, machine tools, and most recently, semiconductors—have suffered the adverse effects of foreign competition. Many of the afflicted firms are desperately trying to play a game of catch-up. Some companies have joined forces with organized labor in an anxious bid to stem the tide of massive erosion of industrial jobs, and are lobbying for various forms of protection (including import quotas and local-content laws) as the last line of defense against foreign competition.

Part I of this book is both an introduction to, and an overview of, the materials presented in Parts II and III. It begins with an explanation of why automated manufacturing is so critical in today's emerging global economy. This is followed by a summary of the underlying reasons that have caused the United States to fall behind Japan and other countries in the adoption of AMES (in-depth analysis of these reasons is presented in Part II). Chapter 3 in Part I proposes certain measures to stem the decline in U.S. competitiveness that has resulted from the slower rate of introduction of automated manufacturing systems relative to foreign competitors. Chapter 3 should be read as an epilogue to the main content of this book. For the sake of convenience it is included in Part I, which serves as an executive summary for the busy policymaker.

p. 329 Survey of Robots

Part II provides a detailed analysis of factors that determine the rate of adoption of AMES in each of the country areas, including analysis of government industrial policies and industry structures and corporate strategies, and user industry demand and management characteristics. (For a synoptic overview of the material presented in Part II, see Table 3 in Chapter 2, "Determinants of AMES Performance: Japan, Europe, and the United States".

Part III contains profiles of thirty-three AMES producers in Japan, Western Europe, and the United States. These profiles provide insights into company product lines, marketing and production strategies, international joint ventures, backward linkages to component suppliers, and forward links (marketing and technical assistance) to equipment and system users. These profiles should prove particularly useful to U.S. equipment manufacturers seeking insights into the competition they face in U.S. and world markets.

For U.S. government policymakers the company profiles will help to explain the industrial management gap between the United States and Japan. For researchers in this field, the book provides a comprehensive analysis of the factors determining the different rates of introduction and diffusion of AMES products in the United States, Japan, and selected Western European countries. This analysis is presented in terms of user-demand factors, producer-supply factors, and elements in the national economic environment that affect both producers and users. Part III is rich in case materials which can be used in industrial management and international marketing courses.

ACKNOWLEDGEMENTS

I wish to thank the following people for their contributions in researching and preparing background material for this book: Leonard Lynn, Bruce Dickson and Kimberly McKeon. I am especially grateful to Robin Roark, who had a major role in writing, editing and analysis from the inception of the project to its final stages of production. I also want to thank the countless persons in business, government, and trade associations who provided much useful information and advice during interviews in the United States, Japan and Western Europe. Finally, I thank Anne Olmstead for her perseverance in word processing the manuscript and Mara Baranson for editorial and research assistance.

Jack Baranson

TABLE OF CONTENTS

TABLES AND FIGURES

GLOSSARY OF ABBREVIATIONS AND ACRONYMS

AMES - Automated Equipment and Systems
AMF - American Machine and Foundry Corporation
APT - Automatic Programmed Tools
BMFT - Federal Ministry for Research and Technology (West Germany)
CAD - Computer Aided Design
CAE - Computer Aided Engineering
CAM - Computer Aided Manufacturing
CEO - Chief Executive Officer
CFC - Compaignie Francaise des Convoyeurs
CIM - Computer Integrated Manufacturing
CODIS - Committee for the Development of Strategic Industries (France)
COGENT - Cooperative Generic Technology Program
Condec - Consolidated Diesel Engine Company
DFG - German Research Society
DFVIR - German Institute for Aerospace Research and Experimentation
DK - Dainichi Kiko
EPA - Economic Planning Agency
FMS - Flexible Manufacturing Systems
GE - General Electric
GFK - Association for Fundamental Technology (West Germany)
GM - General Motors
GMF - General Motors and Fujitsu Fanuc
GNP - Gross National Product
IBM - International Business Machines
ICAM - Integrated Computer Aided Manufacturing
IC - Integrated Circuits
IITB - Institute for Data Processing in Technology and Biology (West Germany)
IPA - Institute for Production and Automation (West Germany)
JAROL - Japan Robot Lease
JIRA - Japanese Industrial Robot Association
KHI - Kawasaki Heavy Industries
MHI - Mitsubishi Heavy Industries
M.I.T. - Massachusetts Institute of Technology
MITI - Ministry of International Trade and Industry (Japan)
MOF - Ministry of Finance (Japan)
MTP - Manufacturing Technology Program
NBS - National Bureau of Standards

NC - Numerically Controlled
NCOP - National Center on Productivity
NSF - National Science Foundation
OECD - Organization for Economic Cooperation and Development
OEM - Original Equipment Manufacture
ONR - Office of Naval Research
OSHA - Occupational Safety and Health Administration
OTC - Osaka Transformer Company
QC - Quality Control
QWL - Quality of Work Life
RAM - Random Access Memory
R&D - Research and Development
R, D&E - Research, Design and Engineering
RIA - Robot Institute of America
ROI - Return on Investment
STU - Swedish Board of Technical Development
UB - Ungermann-Bass Group
UEW - United Electrical, Radio, and Machinery Workers
VLSI - Very Large Scale Integration

PART I
DIAGNOSIS AND PRESCRIPTION

CHAPTER I
WHY AUTOMATED MANUFACTURING
IS IMPORTANT

The Importance of Automated Manufacturing
Industries in Today's Global Economy

The vast new world of industrial automation which has been emerging over the past decade lies at the cutting edge of international competitiveness in today's world economy. International competitiveness has come to depend upon flexibility and versatility in responding to changes in the world economy and upon sustained capabilities to produce in volume increasingly sophisticated and technologically demanding products for global markets at competitive costs. Included here are the use of robots, numerically controlled machines and servo-mechanisms, and what has come to be known as CAD/CAM (computer-aided design and computer-aided manufacturing). At the leading edge of these developments are flexible manufacturing systems (FMS), which by incorporating the aforementioned elements, have revolutionized traditional industrial management practices and efficiency standards.

The new FMS and CAD/CAM systems can produce a much wider range of product variations at relatively low volumes and still be cost-effective, maintain high technical standards, and most importantly, respond to the more rapid and frequent changes in product design and production parameters that are now typical of world markets. The new industrial automation also has another competitive advantage over conventional high-volume manufacturing systems based upon extended product life-cycles and using expensive special-purpose equipment. It avoids the high risks of locking into product designs and capital amortization schedules, which may be rendered obsolete by changes in market demand and shifts in competitive cost structures (as has been the case in the disastrous decline of U.S. automotive market shares in the face of Japanese competition).

It is becoming increasingly evident that *future competitiveness in world industrial markets will depend, to a significant degree, upon a nation's ability and determination to develop and install these new industrial systems as rapidly and extensively as world market competitive conditions change and domestic (economic or political) conditions encourage or allow.* For example, in the short-run, new industrial automation may threaten older generation industrial employment, and there are political and commercial forces willing to resort to trade-restrictive policies to protect jobs and corporate earnings in obsolescent industries. Foreign industrial competitors that are able to maintain high growth and full employment are in a much better political and economic position to adapt the new methods and move to the new industrial modes of operation, particularly if they have effective labor market adjustment mechanisms and accommodating labor-management relations.

In the race for industrial automation, the U.S. economy now faces formidable competition, particularly from Japan, but also from other Western European economies. In the robotics field, Japan is far ahead both in terms of production and installation of these new systems. A significant point is that the *demand* for these systems is apparently much stronger and more extensive in Japan than it is in the United States. This has profound implications for the international competitive position of U.S. industry in the 1980's. The U.S. economy's lagging response to the new industrial systems also has a strong bearing upon the issues of lagging productivity, re-industrialization of declining industries, and the general problem of maintaining innovative dynamics in a modern industrial economy.

Unique Characteristics of AMES Industries*

1.   The new AMES Automated Manufacturing Equipment and Systems industrial complexes, including computer-controlled machining centers and integrated flexible manufacturing systems, imply *cost-efficient, lower-volume* production, coupled with both *time* and *product-mix flexibility*. As a consequence, the *tradeoffs* between added *versatility/flexibility* and *low unit costs* need to be factored into manufacturing and marketing strategies.

2.   These new systems leave *narrow margins for error* at various stages of production (including supplier industries). Product design, quality control procedures, and automated processes become a *closed-loop system*, which necessitates management's combined concern for minimizing production costs, while maintaining *stringent quality control standards* (near-zero defects).

3.   The new AMES complexes require deep-seated *adjustments* in *man-machine, machine-system* relationships. As a consequence, *management and labor motivations and attitudes* are critical to optimal acceptance and utilization of systems.

4.   The potential efficiencies of the new machining centers and totally

---

*See Table 1 for overview.

TABLE 1

AMES CHARACTERISTICS AND CONTINGENT IMPLICATIONS

CHARACTERISTICS (Implicit in AMES)	CONTINGENT IMPLICATIONS (To Maximize Potential AMES Benefits)
Potential Benefits	
- Increased product-mix flexibility	- Capital budgeting that factors in versatility/flexibility
- Increased versatility in adjusting production processes	- Manufacturing has to become an integrated part of corporate strategy
- Extensive opportunities for cost-efficient batch production	- Stringent quality controls (near zero defect)'in manufacturing systems
Requirements	
- Closed loop linkages from design to production and customer servicing	- Labor/management attitudes and motivations favorable to AMES
- Deep-seated adjustments in man-machine relationships	- Effective producer/user forward linkages
- Intricate production engineering of equipment and systems with narrow margin for error	- Effective producer/component supplier backward linkages
- Longer-payback, higher-risk investment in R&D and capital goods	- Management with higher risk propensities and long-term perspectives on corporate returns
	- Reinforcing government-industry relations, supportive financial structures, favorable general economic environment

integrated manufacturing centers lie in their *abilities to respond to market changes rapidly and cost effectively*. But in order to realize these potentials, it is necessary that *manufacturing become an integrated part of overall corporate planning*, and this may imply profound changes in corporate management and practice.

5.   The new systems *require intricate production engineering* (sensors, computers, programs, tools and fixtures, material handling systems, engineering applications, and software). Ultimately, componentry may have to be redesigned and overall systems *tailored to changing and diversified industrial needs*. The foregoing depends primarily upon strong and effective *user-vendor relationships* and complementarities in design-engineering and in run-in and maintenance of systems.

6.   Expenditures on AMES capital goods involve substantial investment outlays and risks and *require longer-term payback perspectives*. As a consequence, the *risk propensities of industrial managers* and corporate commitment to *long-term growth and technological development* are critical. Also important is the impact of *government policies and regulations* upon capital markets, bank lending, and corporate risk management.

Factors Influencing National Rates of Development of Automated Manufacturing Equipment Industries

1.   The National Environment

General economic conditions (including rates and levels of economic growth, employment, interest, wages and inflation) impact upon the demand for, and supply of, AMES products as well as their cost and pricing. Economic recession and inflationary pressures that induce deflationary measures combined with high interest rates depress demand for capital goods in general and AMES products in particular. AMES demand is also influenced by the cost of AMES products relative to the level of industrial wages. The size of the internal market, coupled with the exposure of producers to competitive forces, influences the range and price of AMES products offered to the market. Competitive forces also influence effective demand for technological upgrading among user industries.

Financial structures (including the availability of industrial capital, loan terms, leasing arrangements, and the risk perspectives of suppliers of equity and debt capital) impact upon AMES products' "demand-pull" and "supply-push". Financial resources are needed to finance research design and equipment expenditures and new or improved plant and equipment, as well as to reinforce effective demand for AMES capital expenditures. (See also reference to export credits below.)

Government policies in support of innovation may include tax measures to reduce purchase costs of AMES products, cost sharing of high risks in the development of AMES products, government procurement to reinforce demand when AMES products are first introduced into the market place, and education and training programs to increase the supply of needed technical and engineering manpower. Special programs to assist small-to-medium size industry help to reinforce critical component supplier industries and innovative segments of the economy that have special difficulties obtaining required financial resources. Commercial and trade policies have an indirect effect upon internal competition among AMES producers or upon their competitiveness abroad. Protectionist measures depress domestic demand for the technological upgrading implicit in AMES products. Government-backed export credits reinforce international sales of AMES products.

2. AMES Producers

Commercialization capabilities of AMES producers are a function of corporate management and organization at the enterprise level combined with overall industry structure at the sector or national level.

At the *enterprise level*, factors contributing to commercial thrust include the size of the firm and its ability to mobilize financial and manpower resources; its *forward "outreach"* to user firms, *backward "linkages"* to component supplier industries; and to *previous experience* relevant to the engineering, production and marketing of AMES products. Other determinants of commercialization effectiveness include "product spread" (the range of products and components produced in-house and marketed externally) and the astuteness of corporate strategies in securing market shares domestically and internationally. Cost effectiveness and price competitiveness are intricately linked to these foregoing factors.

Forward outreach implies technical and marketing capabilities to adapt AMES product designs to user needs (including the design of complete manufacturing or processing systems) and to provide ancillary trouble-shooting and maintenance services.

Backward linkages to component and parts suppliers covers a broad and diversified array of components and parts required for computer-controlled machine tools, robotics, and integrated machining centers or entire flexible manufacturing systems. This includes intricate mechanical devices, electronic sensing and control devices, and software programming systems. Producer firms may have to provide technical and financial support to complement supplier capabilities and resources.

Previous experience of AMES producers in applicable industrial fields and relating to the design, engineering, or manufacturing of AMES products or components contribute to the commercialization, marketing and servicing of comparable products. Because of the extensively diversified nature of AMES products, international joint ventures are especially advantageous in securing market shares on a global scale. Through joint ventures rapid development of market shares is possible as a result of achievable complementarities in product lines, production, functions (engineering, manufacturing, marketing) and resources (financial or technical).

AMES sector *structural characteristics* that impact upon the rate of commercialization include a) the number of producers in the market firms; b) the relative size of producer firms; c) the size, range, and degree of sophistication of domestic-user firms; d) types of market segmentation (degree of dependence upon export markets and degree of concentration upon particular niches in domestic or world markets); and e) the degree of competition to which domestic industry is exposed (as influenced by the foregoing characteristics).

3. AMES Users

The *organization and management of enterprises* which acquire AMES products (i.e., users) is a strong determinant in the commercialization rate. "Demand pull" for AMES products is heavily influenced by industrial management's approach to *capital budgeting* policies and the place of manufacturing in *strategic planning* by individual companies. A narrow view focusing upon potential cost savings from the introduction of discrete segments of AMES will limit demand for AMES products.

Potential demand is much greater in corporate strategies that look toward the ultimate advantage of integrated, flexible manufacturing systems in competitive global markets. Similarly an emphasis upon profits in the short-term rather than long-term technological growth leading to expansion of market shares will curb demand for AMES products.

Capital budgeting policies (and the propensity to invest in AMES systems) are also heavily influenced by the previously mentioned environmental factors affecting the availability and terms of credit and leasing arrangements.

"Bottom-up" management that relegates manufacturing to a sub-optimizing level is less likely to move rapidly and extensively into AMES product areas than are firms with "top-down" management policies where manufacturing is an integral and strategic component of overall, long-term corporate planning. The relative prestige and pay of manufacturing managers is relevant in this regard. Rates of commercialization are also heavily influenced by corporate management's "risk propensity"—i.e., willingness to replace existing equipment and systems with the newer generation of AMES products.

The introduction of AMES systems is also favored by industrial management policies that *emphasize "moving down the learning curve"* (i.e., reducing costs through progressive increases in the volume of production) especially when coupled with corporate emphasis upon high-product quality and reliability (i.e., approaching zero defect) and heavy weighting of flexibility in responding to market changes.

The *attitudes of both management and labor* toward improving company performance and perceptions of their respective stakes in company success, have a strong bearing upon the willingness to accept the introduction of AMES systems and product segments. In this regard, the linkage of wages and bonus systems to company performance and measures which minimize the threat of job loss also reinforce the rate of introduction of AMES systems.

Finally, an important determinant of AMES demand is the comparative *sector structures* of AMES-using industries. The intensity of competition among producers (from both domestic and foreign sources) impacts upon the need to improve industrial competition (including introduction of AMES) to retain market shares. Protectionist measures stultify demand for AMES products.

CHAPTER II
WHY THE U.S. IS
FALLING BEHIND OTHER NATIONS

The Japanese industry now has a significant lead on both American and European producers in terms of global market shares. This is due in part to the relative high rate of demand in the sizable Japanese market, which Japanese producers have largely preempted. Even though Japan now exports only a small percentage of the AMES products it manufactures (less than 5 percent), the outlook would seem to be for a considerable expansion of global market shares. The reasons for this are explained in what follows. (See Table 3 for overview.)

User Demand-Pull Determinants

AMES user demand is considerably stronger in Japan than in the United States—a ratio of almost 1:3 when measured in terms of robot sales as a percent of industrial workers (see Table 2). Sweden is ahead of the United States in terms of robotics in use by industry, 1.6:1. The factors contributing to these leads are analyzed in what follows.

General Economic Conditions

In the United States, stagnating economic conditions, declining corporate profits, high levels of industrial unemployment, and inflationary pressures have depressed demand for capital expansion in general and AMES systems in particular. Increased competition of industrial imports in a broadening range of products both from Japan and Western Europe, as well as from low-wage, newly industrializing countries (Brazil and Korea, e.g.), has spurred demand for AMES as a matter of corporate survival (e.g., the automotive industry). Conversely, another reaction to import competition has been rising protectionism, which buffers the import of foreign competition and dampens demand for the technological upgrading implicit in AMES investments.

In Japan, economic conditions have tended to reinforce AMES demand: a higher real annual rate of economic growth over the past decade, unemployment at the two to three percent level, and lower real interest rates (until very recently) to industrial lenders, particularly in favored industries such as robotics. In Europe, declining growth rates, rising unemployment and inflationary pressures on interest rates undermines AMES demand. Until the

TABLE 2

COMPARATIVE DENSITIES OF ROBOT POPULATIONS

Country	Number*	Percent Distribution	Ratio of Robots per 10,000 Workers
Sweden	1,200	9.4	11.2
Japan	6,000	46.7	4.4
United States	3,500	27.3	1.6
West Germany	1,133	8.8	1.0
Italy	400	3.1	0.5
United Kingdom	371	2.9	0.4
France	200	1.6	0.3
TOTAL	12,804		

*As defined by the British Robot Association

SOURCE: Organization for Economic Cooperation and Development, *The Impact of Automated Manufacturing Equipment on the Manufacturing Industries of Member Countries,* 1982.

recent economic downturn, shortages of industrial labor and, in particular, factory skills have spurred AMES demand in Sweden and West Germany especially.

Government-Industry Relations

The introduction of AMES systems generally involves profound changes in industrial organization and management and high financial risk. Consequently, government industry and trade policies and regulatory functions impacting upon risk management are critical.

There are wide differences as to the proper role of government in managing technological change.

In contrast to U.S. practice, Japan and several European governments regard the management of technological change as critical to maintaining the international competitiveness of their national industries. In Europe, reinforcing industrial and trade policies, tax incentives, R&D subvention and government support are strongest in France, weakest in Sweden.

In Japan, the role of government in regard to the foregoing has been purposeful, positive, and pervasive. The principal aims of Japanese Government policy and action have been to reduce private sector risk (technical, market, and general economic); to reinforce the competitive position of Japanese industry internationally; and to do this with a rational, consistent and predictable set of signals and regulations. Particular attention has been focused upon AMES industries as high-impact contributors to the technological upgrading of other Japanese industries. AMES industries also are viewed as a key element in the long-term adjustment toward knowledge-intensive industries to fit the projected population profile of future shortages of a highly educated labor force and shortages in a wide range of industrial skills.

A major aim of Japanese "visions" of industrial sector development is to reduce private sector risk by providing a consensus and framework within which individual firms can develop industrial plans and invest accordingly. Distinctions are drawn between "sunrise" industries (now including robotics) whose development are encouraged, and "sunset" industries that are given phase-out treatment. Government policies and the strategic framework are reinforced by specific measures and regulatory functions, orchestrated and implemented principally by the Ministry of International Trade and Industry (MITI) and the Ministry of Finance (MOF). These include special tax credits and accelerated depreciation on AMES investments and RD&E expenditures, export subsidies, R&D grants, reduced interest rates on AMES investment and leasing, and special aid to small and medium size enterprise. Antitrust laws are liberally interpreted to permit mergers and joint enterprise activities in research and marketing that will enhance international competitiveness. The Japanese Government's indirect role in capital markets for the purchase of AMES products is also substantial. (See below.)

U.S. Government policy, practice, and philosophy stand in marked contrast to the Japanese experience. There is no purposeful or focused effort to favor particular industries, nor to reinforce the *technology* factor in *international competitiveness*. On the contrary, in many areas that impact upon AMES development, government-industry relations are either adversarial (antitrust, e.g.) or ultimately counterproductive (e.g., protectionist measures that undermine demand for technological upgrading).

The U.S. philosophy has been to rely upon the competition of the market place to allocate capital investment resources and to generate technological innovation. This philosophy has severely limited the U.S. Government's role in risk management on the AMES user-demand side and enhancement of the international competitiveness of AMES industries on the producer side.

The U.S. Government is indirectly supportive of AMES technological development through across-the-board financing of basic research and through defense and space-related applied and development research and prototype development. But beyond this limited area of R&D support and related defense-space procurement, there is no specific policy to influence the flow of capital resources into AMES industries or to diminish involved private-sector risks.

Capital Markets and Debt Financing

Several factors contribute to the high investment rate in AMES capital goods in Japan.

The Japanese Government controls levels of bank lending reserves and provides guidelines on allocation of capital expansion funds. Robotics and related AMES products, are designated priority lending areas. High levels of private savings along with government policies which favor lending to "sunrise" industries assure a plentiful supply of debt capital at low interest rates. The latter is especially important to small-to-medium size enterprise that might not otherwise have access to loan capital. (In some areas, the high mortgage value of land provides collateral to purchase AMES for small factories.)

For large companies, the willingness of commercial banks to extend highly-leveraged debt to private enterprise means low overall capitalization costs. Equity investors are accustomed to long-range capital gains and do not insist on short-term dividend payments. Corporate reserves are shielded from tax action, and this adds to funds available for reinvestment in capital goods. All of these factors combine to add to the pool of financial resources for capital expansion and reinforce the abilities of Japanese enterprise to take long-term technological and marketing risks.

U.S. enterprise, both small and large, has a much more difficult task of financing AMES capital expansion. To begin with, U.S. banks will not allow the levels of debt leveraging tolerated in Japan. Second, equity financing is heavily dependent upon retained corporate earnings (after taxes and dividend payments) and upon stock market investments. Third, the concern of American investors in quarterly earnings puts pressure on corporate managers to produce earnings in the short run, and this undermines willingness to make longer term investments and take the higher risks associated with the AMES capital goods. In Europe, on the other hand, capital markets in France are strongly reinforced by government interventions and tax sheltering of capital expansion funds. Swedish government intervention is limited, and Germany is mildly interventionist.

Industrial Organization and Management

Several factors in the organization, strategies and value systems of Japanese corporate management are supportive of a high rate of AMES acceptance. Fundamental is the global view Japanese firms have toward marketing and production and rapid movement down the "learning curve." This implies low-cost, quality-controlled volume production with sufficient variations in product design and performance characteristics to meet diversified and changing international market demands—for which the new reprogrammable manufacturing equipment and systems are particularly suited.

Receptivity to the introduction of AMES into factory systems is further reinforced by managers that take a long-term and comprehensive view toward capital investments which not only considers cost savings in labor, materials and space, but more significantly the broader *strategic* implications of increased flexibility and versatility in designing and producing for shifting demands in world markets. This stands in marked contrast to a much narrower view of capital budgeting generally taken by U.S. firms (see below). The emphasis on long-term growth also predisposes managers to the technological adjustments implicit in the introduction of AMES systems.

The foregoing corporate strategies are reinforced by intensive competition among large numbers of enterprises within each subsection and by *labor and management attitudes and relationships* within individual enterprises. In the latter area, several factors contribute to labor and management's stake in company performance: the wage bonus system linked to productivity gains and corporate earnings, life-time employment for a third or more of the labor force, and labor unions organized at the company level.

The broad experience of Japanese industrial managers and the continuous upgrading of factory labor skills are conducive to an appreciation for, and an acceptance of, newly introduced AMES systems. The technical absorptive capabilities of user firms are also greatly enhanced by management and labor attuned to participating in quality-control circles and related activities to improve factory efficiency on an on-going basis. Many firms have themselves designed and manufactured AMES for use in their own factories.

Enterprise management, organization, and strategies in the United States are less favorable to the introduction of AMES systems than in Japan in several important respects. To begin with, there is a general tendency among U.S. corporate management to be concerned with quarterly earnings rather than long-term growth and technological development. This is in large part conditioned by investors that demand quarterly dividends and low debt leveraging. American management has also shown a marked risk-aversion toward technological upgrading of U.S. plants and equipment. For example, in the consumer electronics and automotive industries, companies have preferred to move production offshore to low-wage countries, rather than redesign and re-engineer (automate) industrial systems.[1]

A third inhibiting factor to the introduction of AMES systems is that manufacturing is not a *strategic* component in long-range planning. One reason for this is that compared to Japanese personnel at the plant management level, American managers tend to be narrowly trained specialists (who are unlikely to have a "strategic" viewpoint) with little corporate experience outside the manufacturing field, which incidentally is lower in prestige and pay than the corporate functions of finance and marketing. As a consequence, corporate decisions on new investments in plants and equipment are viewed narrowly in terms of cost minimization. This is particularly critical in capital budgeting decisions involving AMES investments, where the most significant potential gains lie in a) increased *flexibility* to respond to changes in consumer demands and emergency competition; and b) increased *versatility* in meeting diversified market demands. This relegation of manufacturing to a sub-optimizing level and the failure to think *globally* and *strategically* are formidable deterrents to the widespread application of AMES opportunities. Another deterrent is that too few American firms are committed to "moving down the learning curve;" i.e., product development and production engineering strategies that aim at extensive development of markets, rapid buildup of production volumes, and related pricing strategies that are aimed at capturing significant market shares early in the product cycle.[2]

Finally, U.S. labor and management are not nearly as receptive to AMES introductions as are Japanese management and labor. Few American firms link wage bonuses to company performance; the vast majority of American labor now feels already threatened by increased automation (high unemployment being a strong contributing factor); labor unions are organized by industry or professions (and this does not make for *company* loyalties); and narrowly trained managers are not particularly receptive to new, exotic systems that threaten to displace their traditional skills and experience advantage.

In Europe, Swedish industrial management is most receptive to AMES introduction, followed by West Germany; it is most conservative in France. Co-management arrangement in Sweden and Germany make labor and managers more receptive to AMES than in France. French labor politically is stronger than U.S./Japan and is therefore able to exert protectionist pressures, and unemployment relief.

The technical absorptive capabilities of user firms also contribute to user-demand pull. In Japan, there is high technical absorptive capability among user firms (including effective Quality Control—QC circles). Many users have developed and fabricated their own AMES. This stands in contrast to U.S. firms who are highly dependent upon producers for design, manufacturing, installation and maintenance of AMES, except for a few large firms. This is due in part to the fact that technical adaptive engineering personnel is thin at user end. In Europe, Swedish and German user firms work more closely with producers than do French enterprises in the design, production, installation and maintenance of AMES.

Finally, the competitive structure of user demand industries is another important determinant of user demand. In Japan, intensive competition among a broad range and a large number of domestic producers prevails in most user sectors. Foreign competition is buffered by difficulties encountered in penetrating Japanese markets. In the United States, certain user industries (auto, consumer electronics, and steel) are buffered from foreign competition by import quotas (thereby dulling AMES demand). In Europe, protectionist forces are strong in France, much weaker in Germany and weakest in Sweden (where demand for AMES is most intensive).

Producer Supply-Push Determinants

Industrial Organization and Management

At the *enterprise level*, Japanese AMES producers have very strong forward linkages to user firms, as well as backward linkages to component suppliers. Forward linkages are strengthened by aggressive customer servicing teams that combine design engineering and marketing skills, supplemented by the high level of technical absorptive capabilities that is typical of Japanese user firms. Many Japanese producers that are now in the market designed and built AMES products initially for their own use; this has the added competitive advantage of entering the market with AMES products that have been debugged in the producer's own factory. As a consequence, many Japanese firms are now able to offer completely integrated turn-key factory systems to manufacture and assemble items such as wristwatches and ballpoint pens. In Europe, forward linkages are strong in Sweden and Germany, less so in France. A few firms are experienced in designing/fabricating AMES for their own use.

The *user-vendor relationship* in the United States is much weaker than in Japan in several significant respects. To begin with, few American firms have developed customer servicing to anywhere near the extent or depth of Japanese AMES suppliers (exceptions are U.S. firms such as GCA and Automatix). Most American firms do not have the broad base of technical people (particularly below the university-trained engineer level) to assist user firms in the task of trouble-shooting and maintenance of AMES after installation. Secondly, except for the larger-sized user firms, such as General Electric and General Motors, most U.S. firms rely heavily upon suppliers to adapt industrial equipment design to their individual needs, and few have even ventured into the area of designing their own machine tools. The implication here is that most U.S. firms have limited technological knowhow for working with AMES suppliers, either to analyze their needs or adapt designs to their individual requirements.

Japanese *component supplier industries* generally are highly responsive to OEM (original equipment manufacture) requirements in terms of quality (to near-zero-defects), cost effectiveness, and rapid adjustment to component design changes. This is especially important in the sophisticated componentry required for many AMES products. The OEM-supplier relationship also is highly symbiotic, in that OEM's often provide small-to-medium size component suppliers with financial resources and technical support services and personnel. In Europe, backward linkages to suppliers are strong in Germany, less so in France and Sweden.

Here again, U.S. backward linkages to component suppliers are not nearly as supportive as in Japan. The one exception is in the design and engineering of electronic sensor and control devices and in software design and engineering—areas where U.S. firms have a decided comparative advantage. Most AMES producers in Japan are deeply concerned about potential deficiencies in this area and are worried about being cut off from U.S. technology sources.

Sector Structure

In sheer numbers and in AMES product-range, Japanese firms are well ahead of both their U.S. and European competitors: approximately 150 in Japan as compared to 50 in the United States and 40 in Europe. An indication of product spread is given in Table 4, as reflected in the price ranges of national industry. Domestic competition among AMES producers is very intense, and Japanese user enterprises are both exacting in demands and forthcoming in their technical capabilities to know what they want and to participate in adaptation and run-in of new equipment and systems. By way of contrast, French firms are buffered from foreign competition by preferential credit to French suppliers and by other "buy-national" measures.

Less than 5 percent of AMES output in Japan is now exported, but supplier industries are now reaching saturation points in domestic markets and are projecting 20 to 30 percent export targets over the next decade. Aggressive international marketing networks, coupled with strong financial backing (including special export credits), place Japanese firms in a highly competitive position to take on world markets, as they have in other industries in previous years.

In developing market segmentation, most Japanese firms have entered into one or more joint ventures with both American and European AMES manufacturers for a) cross-licensing of product lines; b) complementarities in product specialization and/or manufacturing know-how (Japanese firms are especially interested in software programming and product design from U.S. sources); and c) enhanced access to foreign markets (either as a hedge against protectionist measures or to reinforce customer servicing through the local national partner). It is interesting to note in this regard, that the set of motivations for entering into joint ventures are somewhat different for U.S. producers. Foreign partners will be used in part for procurement in Japan of effective and high-quality components and parts (and this will inevitably undermine the rate of commercialization in U.S.-based production).

The combination of diversified product spread, strong customer-servicing capabilities, and aggressive international marketing strategies makes for a highly competitive industry in world markets (i.e., for the Japanese rate of commercialization). Japanese AMES producers use global marketing strategies to reinforce overall efforts to achieve volume production and move down the learning curve in order to become even more cost-effective and obtain incremental market shares. Most AMES manufacturers and many of their component suppliers (including small-to-medium size enterprises) have integrated AMES into their factory operations, as part of their efforts to improve cost effectiveness and to maintain the resiliency in their manufacturing operation needed to respond effectively to global marketing opportunities.

By way of contrast, U.S. firms are not nearly as aggressive in international marketing as are the Japanese. Major motives of U.S. firms for entering into international joint ventures are low-cost component procurement from abroad and earnings from marketing foreign imports. In Europe both Swedish and French producers (ASEA and Renault) have aggressive international marketing programs, in part through international joint ventures.

Finally, government support of research and development of commercial applications of AMES products is particularly strong in Japan and France. In the United States, there is special funding for computerized manufacturing in connection with defense procurement requirements, but anti-trust regulations are a *de facto* inhibitor of intra-industry RD&E programs.

Figure 1 provides a diagrammatic presentation of the relative configuration of Japanese, U.S. and European industries. Each of the three diagrams represents an *intuitive* judgment of a) the (horizontal) relative product spread (toward costly and sophisticated products and systems to the right, and toward cheaper and less sophisticated AMES products to the left); b) the relative strength of (vertical) backward linkages to component suppliers; and c) the relative strength of (vertical) forward outreach to users.

TABLE 3

DETERMINANTS OF AMES PERFORMANCE: JAPAN, EUROPE AND THE UNITED STATES

	JAPAN	EUROPE	UNITED STATES
USER-DEMAND PULL			
General Economic Conditions			
Economic Growth	High growth, low level unemployment, low inflation, low interest rates and shortages of industial skills, reinforce AMES demand.	Declining growth rates, rising unemployment and inflationary pressures on interest rates undermines AMES demand.	Stagnating growth rates, high unemployment, inflationary pressure necessitating high interest rates, mild protectionist forces undermine AMES demand. Foreign competition spurs AMES demand for catch up and survival.
Unemployment Levels			
Shortages of Skills			
Inflation			
Interest Rates			
Government-Industry Relations			
Industry-Trade Policies	Strong industry-trade policies to reinforce international competitiveness. Focus on AMES as key industry. Tax measures favor "sunrise" industries; leasing systems; permissive; cooperative R&D programs.	Industry-trade tax incentives, R&D subvention and government support strongest in France, weakest in Sweden. Indirectly support R&D.	Industrial policy to maintain competition through antitrust; government-industry relations adversarial; trade policy not focused on reinforcing U.S. enterprise technological parities. Tax policies do not discriminate in favor of technological lead industries. No specific policies to advance AMES industries, other than defense related procurement.
Regulatory Policies			
Financial Support Mechanisms			

USER-DEMAND PULL	JAPAN	EUROPE	UNITED STATES
Capital Markets and Debt Financing Government Intervention Debt Leveraging Corporate Reserves Investor Expectations	Government guidelines to bank lending channel funds into priority areas. High debt leveraging by banks. Investors attuned to long-term gorwth. Corporate reserves tax sheltered; low dividend payout.	Capital markets strongly reinforced by Government interventions and tax sheltering of capital expansion funds in France. Swedish government intervention limited. Germany mildly interventionist.	Capital expenditures more heavily dependent upon retained savings (after higher taxes and dividend payout). Debt leveraging much lower than Japan's. Stock investors and financial analysts concerned with next quarter earnings.
Industrial Organization and Management Strategic Planning (learning curve) Capital Budgeting Manufacturing in Corporate Strategy	Global view toward marketing and production; drive toward volume production, high market shares, long-term strategic view toward capital investments, technological growth and increased flexibility/versatility in responding to world market demand shifts.	Industrial management most receptive to AMES introduction in Sweden, followed by West Germany; most conservative in France.	User management overconcerned with quarterly earnings, rather than long-term growth and technological development (aspects of risk aversion). Manufacturing not strategic in corporate planning; low in prestige/pay; managers (narrowly-trained specialists) constrain capital budgeting with a strategic view (i.e., rapid movement down learning curve).

TABLE 3—Continued

USER-DEMAND PULL	JAPAN	EUROPE	UNITED STATES
Attitudes of Labor/Management Toward Introduction of AMES Stake in Company Performance Fear of Job Displacement	Labor/Management perceive welfare tied to company performance; broadly experienced managers receptive to AMES.	Co-management arrangement in Sweden and Germany make labor and managers more receptive to AMES introduction than in France. French labor politically stronger than U.S./Japan (able to exert protectionist pressures, and unemployment relief).	Both labor and management feel threatened by new AMES applications. Both fear jobless labor has no stake in improved company performance.
Technical Absorptive Capabilities of User Firms Technical Manpower Past Experience	High technical absorptive capability among user firms (including effective QC circles). Many users have developed and fabricated their own AMES.	Swedish and German user firms work more closely with producers than French enterprises in the design, production, installation and maintenance of AMES.	U.S. firms highly dependent upon producers for design, manufacturing, installation and maintenance of AMES (except for a few large firms). Technical adaptive engineering personnel thin at user end.
Competitive Structure of User Demand Industries Number and Product Range of AMES Producers Competitive Forces in Local Economy	Intensive competition among broad range and large numbers of domestic producers in most user sectors. Foreign competition buffered by difficulties encountered in penetrating Japanese markets.	Protectionist forces in France, much weaker in Germany and weakest in Sweden (where demand for AMES is most intensive).	Certain user industries (auto, consumer electronics, steel) buffered from foreign competition by import quotas (thereby dulling AMES demand).

PRODUCER-SUPPLY PUSH	JAPAN	EUROPE	UNITED STATES
Industrial Organization and Management Forward Outreach to Users / Backward Linkages to Component Suppliers / Previous AMES-related Experience	Strong forward linkages (customer servicing, aggressive marketing internationally); many producers originally developed and built AMES products for own use; relations with component suppliers strong and symbiotic.	Forward linkages strong in in Sweden and Germany, less so in France; a few firms experienced in designing/fabricating AMES for own use; backward linkages to suppliers strong in Germany, less so in France and Sweden.	User-vendor relationship much weaker than in Japan (few firms have strong customer servicing; or international marketing network.) U.S. component suppliers not nearly as responsive to OEM requirements, except in software programs and control/sensor device design.
Sector Structural Characteristics Number, Range, and Size of Firms / Types of Market Segmentation / Competition Facing National Industry	Strong in terms of large numbers and wide range of producers, internal competition among domestic producers, and exacting user demand; future export thrust reinforced by aggressive international marketing capabilities and joint international ventures to assist in market penetration and provide product-design.	European structure weak relative to both Japan/ U.S. in terms of number of competing firms; buffering against foreign competitors (especially in France); a few firms (in Sweden/ France) have aggressive international marketing programs, in part through international joint ventures.	Much smaller number of U.S. firms competing in U.S. market, but competition from Japanese, some European firms emerging; U.S. firms not nearly as aggressive in international marketing as Japanese. Major motives for international joint ventures are low cost component procurement from abroad and earnings from marketing AMES imports.

TABLE 3—Continued

PRODUCER-SUPPLY PUSH	JAPAN	EUROPE	UNITED STATES
Government Support Research and Development Procurement	Joint funding of research design and engineering of next generation of AMES prototypes for commercial applications. Anti-trust does not inhibit joint development programs among competing firms.	French Government highly supportive of AMES RD&E (both commercial and military applications); West Germany to a lesser degree, and Sweden less still.	Special funding for computerized manufacturing in connection with defense procurement requirements. Anti-trust is a defacto inhibitor of intra-industry RD&E programs.

FIGURE 1

AMES PRODUCER INDUSTRIES - COMPARATIVE CONFIGURATIONS:
UNITED STATES, EUROPE AND JAPAN

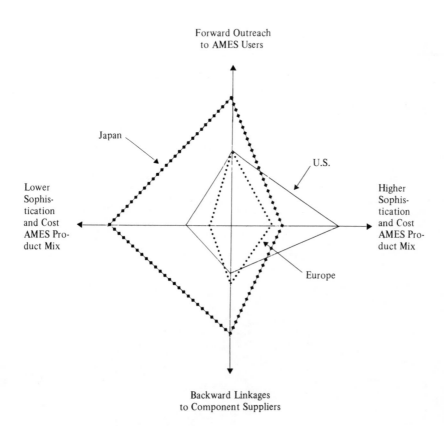

FOOTNOTES

[1]See Jack Baranson, *The Japanese Challenge to U.S. Industry* (Massachusetts: Heath-Lexington, 1981).

[2]For further insights, see Wickham Skinner, *Manufacturing in Corporate Strategy* (New York: John Wiley & Sons, 1978); and Bela Gold, *Robotics, Programmable Automation and Improving Competitiveness* (Ohio: Case Western Reserve University Reprint No. 91, 1981).

CHAPTER III
WHAT CAN BE DONE TO STEM
THE DECLINE IN U.S. COMPETITIVENESS

This chapter is intended as an epiloque to the main body of material presented in the book. It reflects my own judgement and intuition on the underlying causes of our present difficulties, and what I believe we can do to improve our industrial competitive position at home and abroad vis-a-vis Japan.

There is a growing body of opinion among business and government leaders that the root cause of our difficulties in competing with Japan stem from its "unfair trade practices." The thrust of the arguments behind this viewpoint are that (a) the Japanese do not "play fair" in international economic competition; (b) that by means of such practices as "targeting" and "cartelization" they are successfully decimating one U.S. industry after another; (c) that so long as they continue on this course there is no way that the "American system" can either emulate or beat them; and (d) that the only way out of our predicament is to coerce the Japanese into changing their modus operandi by progressively denying them access to our markets.[1]

In my opinion, this viewpoint has several fallacies. To begin with, "targeting" and "cartelization" are scare words for the very effective job that the Japanese have done in moving to progressively higher levels of comparative advantage that are compatible with emerging manpower skills, education, and increasingly effective industrial organization. Consensual management of scarce economic resources represents the Japanese way of looking ahead and adjusting to the evolutionary changes in their own society and in the world economy. It has proven effective in avoiding destructive competition within industry, and adversarial confrontation between industry and government authorities.[2] We are in effect asking the Japanese to limit themselves to our own rules and methods for managing (or failing to manage) economic and social change.

The evidence presented in this book would seem to indicate that a more fundamental and overriding cause of our progressive loss of international competitiveness has been the inadequacy of our industrial response to the Japanese challenge—more specifically, our failure to modernize and automate American industry as rapidly and extensively as has Japan. As indicated in Table 2, Japan now has approximately three times as many installed robots per industrial worker as does U.S. industry. This advantage, in combination with Japan's highly effective industrial policies which operate in harmony with efficient and aggressive industrial management, poses a formidable challenge to American industry.

Another symptom of the underlying causes of our distress is that while Japan is producing the large numbers of technically trained graduates needed to complement the process of industrial automation, we continue to suffer the effects of a shortage of engineers and an acknowledged decline in the quantity and quality of scientific and technical education.[3]

Rather than engage in what may prove to be a fruitless rear-guard action to defend ourselves against Japanese industry by denying access to our markets (and in so doing buffering our industries against the competition that will force much-needed modernization), we should be thinking about public policies and changes in industrial management practices aimed at meeting foreign competition through improved industrial performance. In the vital area of industrial automation, specific recommendations are given below for public policy and improved industrial management of AMES industries on both the supply and demand sides.

1. Government action to reduce management risks on the user-demand side is needed to spur AMES user demand.

User-demand factors are particularly important to the overall growth and expansion of AMES industries. Obviously, a depressed economy with high unemployment levels is not conducive to expansion of demand for AMES. The subsets of interrelated and synergistic factors, delineated under user-demand, impact upon overall risk management[4] within the economy and include capital budgeting at the corporate level, the availability of financing and special tax benefits[5]. The higher rate of growth of AMES demand in Japan, as compared to the United States, is attributable in good part to the role of the Japanese Government in minimizing the risk factor entailed in introducing AMES systems. *The implication for U.S. policy is that it is necessary and possible to influence the direction (and pace) of growth toward particular sectors (and conversely, to ease the phase-out of declining industries) through special tax incentives, loan guarantees, or other special funding measures.*[6]

2. Risk-sharing between government and industry on the producer-supply side is also needed.

On the producer-supply side, the risk-management problem is associated with investments in AMES industries, encompassing both design and engineering and capital investments in plant and equipment. Expenditures on commercializing technological opportunities should be given special tax privileges; and government support of R&D can reinforce the domestic industry, as can military procurement in the early stages of prototype development (see #4 below). *The implication for U.S. public policy is that reinforcing demand-pull is much more important than sustaining supply-push.*

3. Increasing the rate of introduction of AMES may require a special set of industry and trade policies combining (1) and (2) above.

A corollary of both the demand-side and supply-side dimensions cited above is that there is a strong correlation between industrial management at the corporate level and public policies to diminish economic risks, so as to encourage investments in AMES capital goods and RD&E to supply AMES products. *The policy implication is that, as a nation, we may need to consider a combination of industrial and trade policies aimed at reinforcing private initiatives in particular sectors of our economy.* This may include sectors considered vital from a military standpoint or in terms of sustaining employment in industries seriously threatened by international competition. The point is, increased competitiveness would be achieved through technological upgrading (that might include AMES investments) rather than resorting to protectionism. It should be pointed out that from the standpoint of economic growth, it is not justifiable to give the same tax exemption for legal and advertising expenses as for investments in technological upgrading.

4. Reinforcement of AMES industries cannot rely alone on the indirect effects of military expenditures.

In the United States, a strong impetus to AMES industries comes from military procurement—and this is helpful both on the demand-pull side (coming from defense contracts), as well as on the supply-push side (in the form of RD&E). In industries such as military aircraft components, there are likely to be spillover effects to the commercial sector. But this is not true for many other sectors of our economy where revitalization is necessary if they are to continue to compete in the world economy. For example our automotive industry is now trying to play catch-up by robotizing its U.S. production facilities. *The public policy implication is that the United States cannot afford to rely on "trickle-down" effects from military expenditures to maintain technological parities in our industrial economy.*

5. Competitive forces contribute to the level of demand for AMES products and the quality and range of supply; and, conversely, protectionism is inimical to AMES industry vitality.

Reference has been made in our report to the beneficial role of competition in spurring demand for AMES products (to meet foreign competition in particular product sectors such as automotive) and in increasing the range of AMES products offered in these markets. These beneficial effects are bound to be stultified by protectionist measures aimed at maintaining industrial employment.[7] The more advantageous course lies in the direction of industrial and trade policies (as indicated in #3, above).

6. **AMES supplier industries need to improve both their forward outreach to AMES users and their backward linkages to supplier industries.**

Our analysis of the commercial "outreach" of U.S. AMES suppliers, as compared to that of foreign competitors (see Figure 1), points to the need for further improvement on the part of U.S. industrial management in their marketing efforts (to reach both domestic users and foreign markets) and in their backward linkages to supplier industries. *Insofar as corporate policy is concerned, there is much to be learned from the Japanese industry in this regard.*

7. **There is a widespread need for improvement in labor-management relations and in receptivity to AMES.**

There is a pressing need to enhance both labor and management receptivity to technological change and to give labor a broadened stake in technological improvement. The latter is only achievable when low growth rates do not threaten jobs, workers displaced by automation are retrained and/or absorbed elsewhere in the economy, and labor is given its fair share of productivity gains. *From a public policy viewpoint, more attention needs to be paid to the massive problem of retraining and relocating industrial labor.*

8. **Receptivity to AMES also would be enhanced by changes in U.S. management's somewhat narrow views of corporate returns and manufacturing in the corporate strategy.**

Extensive references have been made to certain deficiencies in U.S. industrial management attitudes and practices: overconcern with quarterly earnings, low prestige of manufacturing in overall corporate operations, and the narrow viewpoint toward capital budgeting, resulting in part from narrowly trained specialists lacking the long-range, strategic and global view. *U.S. corporate management needs to face up to these issues, and this will require deep changes in corporate cultures.*[8]

9. **The adverse effects of antitrust regulations on co-operative efforts within the industry to advance AMES technology need to be minimized.**

Reference has been made in the report to the inhibiting effects of antitrust regulations upon intra-industry R&D efforts in the United States. Every effort should be made to minimize these adverse effects by exempting such co-operative efforts within the U.S. economy from antitrust action. (The Export Trading Company Act provides for similar exemptions from antitrust laws.) Otherwise, U.S. firms are often forced to seek technological linkages aboard which sidestep antitrust action and are commercially advantageous, but do not always yield maximum benefit to the U.S. economy.

FOOTNOTES

[1]See for example the series of ads (fifteen in the past year) that have been placed by Motorola in major newspapers and periodicals nationwide. Motorola's credibility is reinforced by its apparent track record in high technology areas and its partially successful effort to penetrate Japanese markets.

[2]For a comprehensive examination of this viewpoint, see Jack Baranson, *The Japanese Challenge to U.S. Industry* (Massachusetts: Heath-Lexington, 1981).

[3]In Japan, all children in grade schools are taught to solve quadratic equations and are introduced to physics and PASCAL (a computer programming language). For an idea of the alarming contrast to U.S. educational standards this represents, see "Can the Schools be Saved?," *Newsweek*, 9 May 1983, pp. 50 ff.

[4]An important dimension of "risk management" is the sharing of risks between public and private sector initiatives. See J. Baranson and H. Malmgren, "Management of Risks in Technology Policies," *Technology and Trade Policy Issues* (Washington, D.C.: Developing World Industry & Technology, Inc., 1982), pp. 164-68.

[5]The current boom in the sale of home and office computers in the face of economic recession is attributable largely to the special tax benefits associated with these purchases.

[6]This has been done in the U.S. housing industry through a combination of loan guarantees, tax incentives, and laws to promote special savings institutions to provide mortgage funds. See also #4, below, with reference to military procurement.

[7]Other distortionary effects of protectionist measures are to countervail the benefits of our antitrust laws and to lead to unhealthy partnerships between U.S. automotive firms and large Japanese companies seeking to breech our trade barriers. See "Toyota's Chevrolet: Protection's Child," *The* (London) *Economist*, 7 May, 1983, pp. 16 ff.

[8]See chapter and verse in Thomas Hout, Michael E. Porter and Eileen Rudden, "How Global Companies Win out," *Harvard Business Review*, September-October 1982, pp. 98-108.

PART II
COUNTRY STUDIES

CHAPTER IV
UNITED STATES

National Environment

Government Industrial Policy

U.S. Government policy has varied from one administration to another, but in general we embrace the philosophy that free market forces should allocate resources and make investment decisions in the private sector. Trade and industry policy are confined largely to maintaining domestic competition (supported by antitrust legislation), and promoting free trade internationally—*not to reinforcing the competitive position of U.S. industry as a whole in world trade* (let alone any particular sector such as AMES). Our national philosophy and policy orientations in regard to the foregoing contrast dramatically with those of Japan. Aside from government support of research and development, national monopoly development and tax and depreciation measures to encourage investment in general, the government-industry relations are adversarial, rather than cooperative, consensual, and orchestrated as they are in Japan.[1]

Tax and Depreciation Measures

These measures are aimed at stimulating research, development and capital investment by industry. Yet real R&D expenditures (adjusted for inflation) in the United States declined steadily throughout the 1970's.[2] Capital investment has declined sharply since the Tax Reform Act of 1976.[3] (Obviously, tax and depreciation measures have not been effective instruments of industrial policy during recent years.)

The Economic Recovery act of 1981 provides for more rapid depreciation of new investments in plant and equipment and increases the size of investment tax credits for industry. Whether these measures will succeed in spite of the current recession (which effectively strangles new capital spending) remains to be seen; they have been characterized as a "new coat of paint" and "more of the same."[4] In fact, the passage of the Tax Equity and Fiscal Responsibility Act of 1982 effectively cancels some 45% of the benefits business received in the 1981 bill.[5]

Legal-Regulatory Measures

The system of antitrust laws in the U.S. was originally intended to deter constraint of trade in the domestic economy. The system has inhibited firms from mergers and/or joint action to share R&D costs, e.g., or to exploit one another's strengths for better position in the international market. The antitrust system hampers the efforts of U.S. firms to meet international competition, since other countries permit or encourage such joint action.

Adversarial confrontation between government and industry, industry and labor, industry and consumers, and among industrial firms themselves drains vital human and financial resources that might otherwise be used to enhance U.S. industrial competitiveness.

AMES-Related Government Policies

Government Support for R&D and Procurement Policies

The U.S. Government has traditionally supported long-range, high-risk research and technology development, leaving the private sector to fund R&D that has good potential for near-term return on investment. Government often provides a market for the fruits of the R&D it sponsors through its procurement policies, thereby minimizing the costs and risks associated with both innovation and commercialization of new technology.

Total U.S. spending on R&D is more than the amount spent by major OECD competitors combined. About 50% of total U.S. R&D expenditure annually is financed by the Federal government, of which approximately 50% is for defense-related projects and another 20% for aerospace projects.

The bias towards military and aerospace projects in government R&D funding and procurement, together with its emphasis on basic research (with only long-term payoffs for industry) means that little government support is available that directly enhances industrial competitiveness and commercialization. Funding for R&D is the only explicit way in which the government supports the spread of AMES and is generally limited to mission-oriented national defense needs.[6]

The military focus of government R&D on manufacturing technology has two detrimental side-effects: a) neglect of generic, systems approaches, limiting the applicability of results to civilian problems and settings; and b) the emphasis on hardware and predominance of physical scientists and engineers on program staffs results in neglect of social and organizational factors in design of manufacturing systems, making user implementation of the technologies developed more difficult.

The Defense Department's Manufacturing Technology Program provides hundreds of millions of dollars annually for purchase of advanced manufacturing equipment. But very little pressure is exerted on its R&D contractors to become cost effective and internationally competitive through commercialization of technological innovations. The program includes capital investment initiatives to encourage modernization of the defense industrial base by: a) increasing program stability and use of multiyear procurement, b) supporting legislative efforts to revise tax and profit policies, c) improving contract incentives, d) increasing direct investment in technology for the private sector. None of the foregoing was taken specifically to foster factory automation nor to improve productivity in the private sector, and their effects on private sector capital investment and commercialization of AMES are uncertain.

AMES-Related Manpower Policies

Commercialization of AMES is hampered by a shortage of qualified personnel on both the producer and user sides. The Federal role in manpower for robotics is representative of its general role in engineering and scientific manpower development. Government programs in this area have been politically controversial, since many people believe that "market forces" should determine the supply and demand of labor without government intervention.

Government efforts to date have been effective in identifying manpower requirements. Programs to increase the supply of engineers, technicians and scientists have been less effective, being hampered by lack of cooperation and coordination with industry, state and local governments, and educational institutions. Measures to improve this situation are contained in the National Engineering and Science Manpower Act under consideration in the House of Representatives (H.R. 5254), but enactment does not seem likely in the foreseeable future.

AMES-Related Trade Policies

Rising protectionism in the United States (including recent measures against steel imports and import quotas on consumer electronics and autos) buffers industry against foreign competition and slows technological upgrading, including adoption of AMES. The National Machine Tool Builders Association (which represents several AMES producers) recently called for protection against rising import penetration (36% at present) on grounds of national security.[7]

Resources spent on protectionism could be better invested in commercialization of AMES, yet political pressures for protection are increasing. Some elements within government realize that the potential threat of competition from automatic factories abroad outweighs the short-term hardships countered by protectionism.[8] Previous governments have begun efforts that could have helped U.S. industry compete through joint R&D on non-military manufacturing technology. The National Center on Productivity (NCOP) of the Nixon administration and the Cooperative Generic Technology Program (COGENT) during the Carter period were both abolished by successor administrations. President Reagan created a Productivity Council in 1981, but the first action of the Council's director was to go to Japan to argue for voluntary restraint in exports to the United States. Clearly, short-term political pressures hamper the ability of U.S. governments to implement medium and long-range programs to encourage much-needed industrial automation.

Government Measures Regarding AMES

As noted above, explicit measures fall largely into the category of support for R&D. Total U.S. government support for R&D on robotics is estimated to be about $18 million (FY1982). The National Science Foundation (NSF) is spending about $4.5 million per year on civilian robotics and related research, distributed largely to universities and non-profit labs in small grants of about $200,000 each. NSF primarily supports basic research centering on improved robot dexterity and sensory perception, higher-order robot programming languages, and computerized manufacturing and assembly.

The National Bureau of Standards (NBS) spends $1.2 million per year on inhouse robotics research, generally focused on robotic applications. An important area of NBS' work is on robot interface standards. Such industry-wide standards could provide users with flexibility in building automated systems and increase user confidence in their quality, thus fostering adoption of AMES. However, the development of standards for advanced automation technology has been slow. Government has not unilaterally set these standards but has worked with industry to voluntarily build a consensus on the standards industry should adopt. While AMES producers abroad have moved to adopt international standards, U.S. industry has been unable to achieve a consensus.

The Office of Naval Research (ONR) distributed $1.7 million in FY1982 to a small number of universities for a mixture of basic and applied robotics research in response to the Navy's needs. NASA spent $2 million in FY1982 on robotics R&D directly related to its mission, and the Veterans Administration averages about $500,000 per year to support research on potential robot applications for the handicapped.

Each of the U.S. military services spends significant amounts ($1-10 million annually) to support mission-oriented basic and applied research and development in robotics (Science and Technology Program). Each of the services also spends between $1 to 5 million a year on procurement to support adoption of AMES by the defense industrial base (Manufacturing Technology Program—MTP). The Air Force in particular has been active in promoting adoption of AMES by the aerospace industry (ICAM Program).[9]

In the past, military funded R&D led directly to development of numerically controlled (NC) machine tools and the automatic programmed tools (APT) language, both of which are now in commercial use worldwide. Doubts about the ability to commercialize the results of military-oriented AMES R&D remain, especially in comparison to the more commercially oriented efforts of foreign governments.

AMES Producers

Overview of Industry Structure

Producers of automated manufacturing equipment and systems belong to an emergent industrial sector that brings together several diverse areas of manufacturing and information technology. The hardware component of AMES involves producers of machine tools, hydraulics, electromechanical and electronic equipment, and sensory devices of several types (optical, electromagnetic, tactile, etc.). Software elements of AMES are produced by firms active in the fields of communications, computers and information science.

Several producers have overlapping activities in more than one of these fields. A few firms may some day be able to integrate all of these elements to provide turnkey flexible manufacturing systems (General Electric and Westinghouse have this potential; GCA has stated an intention to do so). Today, however, the producer industry is dominated by a robot pioneer, Unimation, and a machine tool builder, Cincinnati Milacron.

Early Developments in U.S. Robotics Industry

During the 1940's and 50's, development work was done on most of the elements of industrial robots. U.S. patents were issued for work on articulated arms, remote materials handling, manual "teachability" (in connection with early paint spraying machines), and programmable controls (both punched tape and microprocessor-based types.)[10] By the mid-fifties George Devol was citing many of these patents in his own work on a Programmed Article Transfer System which was patented in 1961 and generally recognized as the first programmable industrial robot.

The Devol patents were bought as a package by Consolidated Diesel Engine Company (Condec) which set up a subsidiary, Unimation, Inc. to exploit them. The first servo-controlled, continuous-path Unimate robot was sold to General Motors in 1961. American Machine and Foundry Corporation (AMF) began to market its non-servo-controlled, point-to-point Versatran robots shortly thereafter.

These early robots were unproven and risky. They were expensive (in the context of the times), costing an average of $25,000 each, and about $4/hour to operate. Their control and feedback technology limited the application of robots to only a few simple jobs. As a result, there were only about 200 robots in the U.S. by 1970, used primarily in jobs that were too hazardous or boring for humans.

This situation changed during the 1970's. Productivity growth in manufacturing slowed at the same time wage rates increased rapidly. The cost of many manufacturing inputs increased as well, due to successive "oil shocks." Manufacturing managers intensified the search for cost-control methods.

The control, flexibility and manipulative capabilities of robots were being improved at the same time. Cincinnati Milacron had developed a system to control several NC machine tools directly from a central computer (thus by-passing tape readers) in the mid-1950's. Building on this experience, Milacron introduced the first minicomputer-controlled robot, the T3 in 1974. Minicomputers were replaced by microprocessor controls during the late 1970's. As CAD and AS/RS systems were developed and refined, their links to CAM were explored, and computer integrated manufacturing (CIM) became a real possibility.

In response to these developments (economic and technological) robot use grew somewhat during the late 1970's. Yet by 1978, the Robot Institute of America (RIA—the principal U.S. trade association for robots) consisted of just 10 manufacturers, 3 accessory makers, 25 users, and 3 research organizations. In that year, some 650 robots were sold (for a dollar value of about $46 million), raising the total of robots installed to approximately 2,500.

Present Structure

Interest in AMES products has grown rapidly since 1978. The increase in actual sales and in the use of robots has been less spectacular, but still significant. It is estimated that 2,100 robots were sold in 1981 (for a total value of $155 million), increasing the base of robots installed to over 5,000. During the past five years of rapid growth, the number of firms supplying robots has grown from a dozen to approximately fifty. During 1982 a half-dozen firms entered (or announced their intention to enter) the U.S. market as suppliers and/or manufacturers of AMES.

This dynamic state of the producer industry is complicated by restrictions on capital spending (caused by the recession) among many potential robot users. A large portion (up to 50%) of the rapid market expansion since 1978 was due to large orders for robots placed by the automobile makers (see AMES Users, below). Such an order volume may not be sustained if the hard-hit automotive sector puts off capital spending in order to weather the recession. This will only delay the growth of some AMES producers; others, especially smaller firms and start-ups less able to absorb losses will drop out of the market.

At present, two firms account for nearly 75% of total U.S. robot sales. Unimation's 1981 sales of $68 million (43% market share) represented over 17% of parent Condec's corporate sales. Cincinnati Milacron, the largest U.S. machine tool manufacturer, had robot sales of $50 million in 1981 (32% market share), representing only about 5% of total corporate revenue.

In 1981, the top eight AMES producers accounted for 97% of an estimated $155 million in U.S. sales.[11] Leaving aside the two industry giants described above, these firms had sales of between $3-9 million apiece. In order of declining market share, they are: ASEA (U.S. subsidiary of the pioneering Swedish firm); Prab Robots (which purchased AMF's Versatron robot line); DeVilbiss (developer of some early fixed-sequence paint spraying robots, now licensing technology from Norway's Trallfa); Copperweld Robotics (created by Copperweld Corporation's purchase of established robotmaker Autoplace); Automatix (largely a systems integrator, stressing control and sensory technology); and Nordson (a manufacturer of surface coating equipment now licensing robot technology from Italy's Basfer to round out its product line).[12]

The total U.S. market for AMES should reach $2-4 billion by 1990. Several very large firms have entered the market (Bendix, General Electric, General Motors, IBM, Westinghouse) or may soon enter (Digital Equipment Corp., Texas Instruments, Textron, United Technologies) in anticipation of the projected growth. The entry of these firms will intensify several trends as the industry matures. The end-user market will become much more diversified. This is already happening as potential users become aware of robotic applications in their industries, and as current robot producers struggle to expand sales in spite of slumping orders from the auto industry. Some of the new entrants will use large numbers of robots internally. All of them have a strong sales presence in other industries, enabling them to effectively broaden the user base.

The entry of large firms will also accelerate the trend towards functional diversity in robot applications. Up to 35% of all robots sold in 1980 were used for automobile spot welding. Functional diversity is increasing not only in response to the "pull" of a broader base of users, but also because large firms can "push" new technology to capitalize on some internal strength (e.g., in a related product, a proprietary innovation, or in engineering a particular application).

IBM, for example, first began R&D on robotics in 1972 in order to find new applications for computers on the factory floor. The robot arm of its new 7535 Manufacturing System (primarily for light assembly) is made by Sankyo Seiki in Japan. The $28,000 robot is controlled by a $5,000 IBM personal computer (and software), reflecting IBM's traditional strength.

As functional applications become increasingly diverse, the producer industry is likely to respond by becoming more segmented. In the area of surface coating, for example, DeVilbiss has been clearly dominant, based on its previous experience in the field. Some firms may choose to limit themselves to a certain market segment (according to application, level of technology, geographical area, price, etc.) as a conscious strategy to establish a defensible niche in the face of increasing competition. Mobot Corporation, for example, has concentrated on selling non-programmable robots to gain a small but relatively secure share (1%) of the U.S. market.

Finding a secure niche will be difficult in the long run as the number of competitors increases in each. DeVilbiss has been joined by Nordson, Binks, Graco, Cybotech, Thermwood, General Electric and General Motors in the area of surface coating robots. There are at least nine U.S. firms competing in the field of arc-welding robots, a rapidly growing application that now accounts for only about 5% of the market.

With increasing competition, prices have decreased. The average base price of an industrial robot was about $70,000 in 1982 (see Table 4). In the short-term, pricing policies may be very aggressive due to weak demand. ASEA, for example, reduced the price of its two most popular U.S. models by 25% at the end of 1981. This was done partly to relieve excess capacity of the parent Swedish company since those models were not produced in the United States at that time. It was also a bid to gain market share, probably at the expense of Cincinnati Milacron's comparable models.

Over the long run, however, several factors will gradually lower the price level:

- U.S. producers are moving along the learning curve in robot production. Cincinnati Milacron, Unimation, Automatrix, ASEA, and Prab have all recently dedicated new, more efficient production facilities to capture economies of scale.

- The price/performance ratio of robots (and of their components such as microprocessors) will improve, especially as larger firms with greater manufacturing experience enter the market.

- Foreign competition will intensify. Robot imports from Japan, less than 5% of U.S. sales in 1981, could account for 25% of the domestic market by 1985. Whether sold directly or under license by U.S. firms, many Japanese robots are priced lower than comparable domestic models.

TABLE 4

PRICE RANGE FOR ROBOTS*

Country	Price Range (US$)
United States	$10,000 - $150,000
Japan	$ 6,825 - $100,000
Europe	
West Germany	$13,200 - $123,200
United Kingdom	$17,400 - $110,400
Sweden	$ 8,860 - $ 90,000
Norway	$ 8,519 - $ 85,175
France	$ 4,000 - $160,000
Italy	$10,000 - $150,000

*RIA definition of robot excludes mechanical transfer devices.

SOURCE: Adapted from Table 3, page 4, *RIA Worldwide Robotics Survey and Directory,* Robot Institute of America, Dearborn, Michigan, 1981.

Falling prices will stabilize over time and some AMES products for certain well-established applications will behave like commodities, as already reflected in the price/performance ratios for auto body spot welding robots. Continued technological innovation in robot materials, controls, drive systems, sensors, architecture, and especially software, together with price, will be the bases of competition through the 1980's.

Corporate Strategies and Outreach

In general, robot manufacturers in the United States have emphasized marketing or research and development over manufacturing in corporate strategy. An industry leader is reported as having a disorganized manufacturing operation, with production organized as in a large job shop and with poor quality control. In earlier years the problem was of less concern since most buyers were fairly sophisticated users.

However, some U.S. producers have strongly emphasized manufacturing and expanded capacity in preparation for future growth. GE and

Westinghouse, among the larger new entrants to the robotics market, have emphasized the relationship of manufacturing and corporate strategy.[13]

Producers that place a high priority on manufacturing can achieve lower costs and higher quality, ultimately leading to greater consumer acceptance and commercialization.

Forward Outreach to Users

Forward outreach involves all the efforts a producer makes to "push" finished products out of his warehouse into the user's factory floor. These efforts include all those functions that make up sales and marketing. Most important of these functions—because AMES represents process technology— is the ability of the producer to understand the user environment and to integrate his equipment into it.

U.S. robot producers have had such a long and close relationship with users in the automotive sector that many neglected user outreach until recently.[14] Unimation has sold nearly 70% of its robots to relatively sophisticated users (capable of doing their own applications engineering) in the auto industry; its efforts to broaden its customer base undoubtedly suffer from the weakness of its system group.[15]

As users become more aware of the systemic nature of computer integrated manufacturing, they will increasingly look for systems suppliers with a good understanding of the factory environment. Systems are expensive, and successful producers will have to finance in-process inventory of other system components as well as their own production. Users of expensive systems will be reluctant to deal with any but the most financially stable producers.

One small producer specializing in turnkey robot systems built around a vision capability is Automatix, Inc. The firm was founded in 1980 with $6 million in venture capital by Philippe Villers, who earlier co-founded Computervision, a producer of CAD systems. Automatix achieved a 2% market share (on sales of $3 million in 1981), perhaps due to the knowledge of the user environment of some key staff people.

The entry of companies such as GE, Westinghouse, IBM and Texas Instruments will add several potentially strong system suppliers to the U.S. industry. All of these firms (and others including Cincinnati Milacron and GCA) have acknowledged the importance of the systems approach to factory automation. Each would probably have a differential system based on an internal strength (e.g., Milacron integrating robots with machine tools). None is likely to become a commercial success in robotics without some strength in the following elements:

- a strong field service organization for customer support;
- continuous and adaptable R&D;

- software enhancements;
- knowledge of the user environment;
- financial strength;
- long-term commitment to the market; and
- vigorous and broad-based marketing.

Backward Linkages to Component Suppliers

The U.S. producer industry is heavy with firms that rely on suppliers for a greater or lessor proportion of components which they assemble into automated manufacturing equipment. A few of these producer-component supplier relationships are long-term, stable, and quality-oriented (as in the case of Prab Robots and Robot Research Corp.)

An increasing number of such relationships will represent internal sourcing of components, by one division of a diversified parent company from another ·(this is the case with GE and Westinghouse). In all cases, part of the responsibility for manufacturing productivity and product quality is effectively transferred from the producer to the component supplier. Emphasis on lowest-cost sourcing of components implies frequent changes of suppliers for many U.S. producers, and sometimes inadequate quality control. The reputation for reliability of some early U.S. robots suffered from component breakdowns.

Product Line "Spread"

Of the top six U.S. robot producers, only one (DeVilbiss) has a product line that encompasses fewer than three different functional applications. Even DeVilbiss, market leader in surface coating robots, is broadening its product line by adapting its paint spraying robot to the requirements of arc welding.

U.S. producers have, in general, emphasized a broad product spread somewhat skewed toward the high range of price and sophistication. The average robot price in the U.S. is $70,000; significantly higher than in Japan and Europe. The proportion of sophisticated robots (i.e., continuous-path, microprocessor controlled and/or sensor equipped) to total robots installed is higher here than elsewhere. Since many users are highly sensitive to price (basing their investment evaluation on payback—see AMES Users, below) the broad and high-priced product line offered by most U.S. producers has inhibited commercialization. Some producers have licensed foreign technology specifically to broaden their product line (see Joint Ventures, below) thereby offering lower-cost robots as well, in some cases. The result may be negative customer reactions, as the forward outreach provided in support of a broad product spread is spread thin.

Joint Ventures, Licensing Agreements and Acquisitions

Several U.S. producers have used one or more of these mechanisms in efforts to broaden product line, gain a manufacturing base, or gain market share. Unimation signed a technology exchange agreement with Kawasaki Aircraft (later absorbed by Kawasaki Heavy Industries-KHI) in 1968. Unimation's initial motive was simply the large royalty payments flowing from the agreement, but within five years, the flow of technology from Kawasaki had become important as well. This agreement continues to be of importance to both parties today.

The only other U.S. producer to aggressively pursue foreign licensing has been Prab, with licencees in Japan, Canada and Europe. Prab's primary motive at this point is to derive fees and royalties from foreign sales of its technology, since it has no internal mechanisms for international sales.

The remainder of U.S. firms listed in Table 5 license foreign technology or simply buy AMES products for resale in order to have something with which to immediately establish a market presence. Some firms, such as GCA, are developing products in-house that will not be market-ready for 3-4 years, and licensed technology provides revenue and experience in the interim. General Electric became a supplier of a dozen different robots (from three licensors) in the space of about two years. In this way, GE was able to gain invaluable experience through testing in its own facilities a variety of products for potential user applications and higher performance standards. Westinghouse, GE's direct competitor, has followed much the same strategy.

The joint ventures between Ransburg and Renault (Cybotech), and General Motors and Fujitsu Fanuc (GM-Fanuc) both have, or will have, production facilities in the U.S. and are thus targeted at the domestic market. GM gains the assistance of low-cost producer (Fanuc) while providing broad knowledge of U.S. applications. Ransburg gains technology and applications experience from Renault, who in turn benefits from a local source of robots for its Wisconsin auto plants (joint venture with American Motors).

Several firms have used acquisitions to enter the AMES market or to expand existing operations. Prab acquired AMF-Versatron (a robotics pioneer) in 1979, thereby expanding its robot line by 50%. GCA acquired PaR Systems (maker of rail and trolley type robots) in 1981, and with the addition of an experienced staff and a line of robots licensed from Japan, GCA became a potentially strong systems producer in about a year's time. GE acquired Calma (a software house) to round out its licensed robot technology. This trend of merger and acquisition in the AMES field is likely to continue. Small software houses and developers of sensory devices are likely targets for larger systems-oriented firms.

TABLE 5

LICENSING AGREEMENTS AND JOINT VENTURES
OF U.S. PRODUCERS

From	To

JAPAN

1.	Dainichi Kiko	GCA
2.	General Motors	Fujitsu Fanuc
3.	Hitachi Ltd.	Automatix
4.	Hitachi Ltd.	General Electric
5.	Komatsu	Westinghouse
6.	Mitsubishi Electric	Westinghouse
7.	Nachi Fujikoshi	Advanced Robotics Corp.
8.	Prab Robots, Inc.	Murata Machinery Ltd.
9.	Sankyo Seiki	IBM
10.	Unimation	Kawasaki, H.I.
11.	Yaskawa Electric	Hobart Brothers
12.	Yaskawa Electric	Machine Intelligence Corp. (MIC)

Notes:
1. Has future provision for technology exchange, joint R&D.
2. This is not a licensing agreement, but rather an equally-owned joint venture in which product and ideas will flow from both participants.
6. This is not a licensing agreement but rather a sale-for-resale type of accord.
10. There is also some flow of product back from Kawasaki, but Unimation has been the dominant force in this relationship.
12. Yaskawa Electric will provide MIC with robots; MIC will provide Yaskawa with vision sensory devices.

ITALY

1.	Basfer	Nordson
2.	DEA	General Electric
3.	Olivetti	Westinghouse

BELGIUM

1.	Prab Robots, Inc.	Fabrique Nationale

Note: Again, an opposite-direction flow.

FINLAND

1.	Unimation	Nokia

Note: An opposite direction flow.

WEST GERMANY

1.	Nimak	United Technologies
2.	Volkswagen	General Electric

FRANCE

1.	Renault	Ransburg

Note: This represents a joint-venture effort, Cybotech.

NORWAY

1.	Trallfa	DeVilbiss

SOURCE: Adapted from *Robotics Newsletter*, No. 8, May 4, 1982, p. 5.

AMES Users

Overview of AMES Demand

In 1979, only about 700 industrial robots were installed in the U.S., amounting to roughly $65 million in sales. In 1981, U.S. industrial robot sales amounted to roughly $150 million, with a cumulative total of some 5,000 robots installed by year's end. This is a small market by the standards of U.S. industry, representing less than one-thousandth of one percent of U.S. GNP in 1981. Yet future sales growth by most estimates will be very strong, 35-40% per year through 1990. In that year, over 30,000 units sold could provide revenues of over $2 billion.

The current recession has certainly slowed robot demand in the U.S. Many AMES producers have reported stagnation or even decreases in new orders during 1982, in contrast to steadily increasing order backlogs for the previous several years. However, a recent informal survey of metal working industries found that roughly 22% of the sampled firms included robot purchases in their capital budgeting plans during the 1982-1984 period.[16] The survey also found that plans for robot purchases were evenly distributed among different industries in the metal-working sector.

Within this small but growing market AMES producers that can marshall their resources in response to identified or predicted demand trends will be most likely to achieve success in commercializing their technology. Trends in robot demand can usefully be examined from two perspectives: 1) according to the functional type of application (e.g., welding, assembly, etc.); and 2) by end-user industrial sector.

Functional Demand

Spot welding was one of the first applications of industrial robots, and is currently the leading one in the U.S., accounting for roughly 40% of all robot applications (see Table 6). This reflects the swift adoption of robot spot welders in the manufacture of auto bodies. While there is still room for growth in spot welding applications, the market will be almost entirely replacement orders by 1985. Spot welding applications are estimated to be only 8-10% of robots in use by 1990.

Arc welding is a more difficult task for robots than spot welding, since it requires some type of seam-trenching capability. Thus, only about 5% of robots in use in 1981 were arc welding, most using rudimentary contact sensors for seam tracking. Demand is strong however, since arc welding is extremely dirty and dangerous for humans. With the development of effective and low-cost vision systems for seam tracking, arc welding should account for roughly 15% of robot applications in 1990.

The second most important application at present is materials handling (including palletizing, parts handling, and machine loading/unloading). Roughly 35% of robots are now used in materials handling applications. Robot loading/unloading can maximize the productivity of NC machine tools, which are already in widespread use. This particular application has grown strongly since 1980 and the trend should continue through the mid-80's. Materials handling applications are projected to account approximately for 30% of robots in use by 1990.

TABLE 6

U.S. ROBOT DEMAND TRENDS BY APPLICATION

Application	Percent 1981	Percent 1990
Spot Welding	35-45	3-13
Arc Welding	5-8	10-20
Materials Handling (including machine loading and unloading)	24-45	25-35
Surface Treatment	5-12	5
Assembly	3-10	25-40
Other	7-10	7-14

SOURCE: Composite of estimates from: 1) L. Conigliaro, "Bullish Days in the Robot Business", *Robotics Age*, Sept./Oct., 1981, p. 6; and 2) R.J. Sanderson, et al., *Industrial Robots*, Tech Tran Corp., Naperville, Illinois, 1982, p. 125.

End-User Demand

The use of industrial robots for assembly tasks should increase dramatically during the 1980's. Currently only about 10% of robots are used in assembly applications, (generally very simple). With the development of improved positioning accuracy, vision, and tactile sensing capabilities (currently the foci of intensive R&D efforts both in the U.S. and abroad), assembly application could account for 35% of robots in use by 1990.

Demand for industrial robots used in surface treatment applications (spray painting, resin coating, etc.) will increase in absolute terms. The share of total robots in use accounted for by surface treatment applications should be approximately the same in 1990 as it was at the end of 1981: between 5% and 10%.

Other applications of industrial robots (e.g., machining, forging, inspection) account for some 10% of robots in use at present, and this will likely remain the same in 1990. The relative share of inspection robots in this "other" category should increase (if effective sensory capabilities become affordable), since user interest is strong.

With regard to robot demand broken down by end-user industrial sector, there is little reliable data available. Many analysts believe that at least 80% of robot orders in the U.S. come from about 20% of total robot users. It is generally acknowledged that the automobile industry accounts for some 50-60% of robots currently in use. Probably one-third of U.S. robots are now used in a single application within the automotive sector—spot welding of auto bodies.

The number of firms using industrial robots is relatively small, in the range of 300 to 400. Metalworking industries other than automotive account for the majority of remainder of robot use. Within the metalworking sector, the casting, foundry, and light to heavy metal fabricating industries all use significant numbers in robots. Demand is increasing in the aerospace, electrical/electronic, and plastics industries, though they account for relatively small numbers of robots at present.

Rationale for AMES Demand

In recent years between 50 and 75 percent of the dollar value of durable goods manufactured in the United States has been "batch-produced," contrary to the popular image of pervasive mass production.[17] Shifts in consumer preferences and growing demand for customized products are forcing U.S. manufacturers to simultaneously increase product variety, quality and efficiency in batch production.

It is widely recognized that industrial robots have the potential to increase productivity, help control product quality, provide flexibility in adapting to product changes, and reduce unit costs, especially in batch manufacturing operations where production volumes are in the middle range and where neither manual labor nor hard automation has a cost advantage.

It has been estimated that currently available robots are capable of performing 15 to 20 percent of the total of 7 million jobs in U.S. manufacturing plants that could someday be robotized (with improved robot capabilities).[18] This represents a theoretical potential of over one million robots that could be utilized at present. Yet robot orders have slowed in 1982;

clearly the weakness of effective demand in the United States has been a brake on commercialization of AMES.

AMES Versus Traditional Technology

Process technology in certain industries (such as steel, chemicals and pulp/paper) has been highly developed over the past thirty years, primarily to achieve the huge economies of scale sought in these sectors. Yet in industries where batch production is important (e.g., metalworking, electronics and furniture), development of process technology has been slow and evolutionary. Automated production technology for batch process industries (including the NC machine tools and dedicated transfer lines adopted by the auto industry) has tended to be costly relative to its benefits, inflexible in terms of product changes, efficient only at high production volumes, and full of the "bugs" inherent in what are often custom-designed new processes.

Suppliers of process technology have offered specialized equipment that often proved difficult to coordinate with other machine tools, or did not span enough of the user's process steps to provide major economies. Faced with process technology inadequate to their needs, batch manufacturers have typically spent money freely for product innovation, but have been more reluctant to invest in process innovation. This reluctance continues today, even though the new automated manufacturing technologies are more and more qualified to meet the needs of batch manufacturing.

Poor performance of some early robots (perhaps in combination with overblown expectations on the part of robot users) in the 1960's and 70's had some part in the continuing reluctance to invest in AMES, yet in a recent survey, 85% of all users stated that their robots either met or exceeded their original expectations of performance. The prime reason for continued reluctance to invest in AMES is no longer inadequate technology, but inappropriate attitudes and techniques on the part of U.S. industrial managers. (See below.)

Impact of Economic Environment Upon Effective Demand

Several negative economic factors have inhibited effective U.S. demand for AMES over the past decade. Low levels of economic growth, stagnant productivity growth, rising unemployment and high interest rates have affected all sectors of U.S. manufacturing to a greater or lesser extent, depressing the general level of demand. Effective demand for AMES in particular has been inhibited because:

1)   AMES costs and risks are high and require a high return;

2) high interest rates and capital shortages require rapid payback and capital investment by many firms is limited to immediate necessity; and

3) U.S. corporate tax laws (even as improved by rapid depreciation, lower tax rates and investment credit deductions) still require a higher ROI than in other countries.

The average programmable industrial robot (RIA definition, types A-D) in the United States costs $65,000, more or less, depending on the type, application, and capabilities. The average installed price is $115,000 including the basic robot, accessories (maintenance and test equipment, grippers, etc.) and installation costs.[19] Thus, even a single robot represents a substantial capital investment for many of the smaller firms engaged in batch manufacturing that could most benefit from robotics.

A typical multi-robot installation in the automotive industry requires an investment of from two to four times the actual robot cost, depending upon the application. In a major body assembly (spot welding) system, the cost of fixturing, welding and material handling equipment will be three to four times the cost of the robots involved.

General Motors' announced goal of utilizing 14,000 robots (in body and parts assembly, machining, stamping, casting and forging operations) by 1990 requires an investment of nearly $1 billion (at the average robot price of $65,000 mentioned above) for robots alone. The total investment required (from two to four times that figure), is large even by GM's standards. The ability of GM and the rest of U.S. industry to invest heavily in AMES in the face of high interest rates, low profitability, capital shortage and a general economic recession is questionable. The dilemma is that investment in AMES is part of the long-term solution to these economic ills.

Automated manufacturing equipment and systems can have substantial costs in addition to the installed cost (e.g., for training, maintenance, and parts inventories). Individual machines can also cause production bottlenecks—if, for example, a plant lacks skilled personnel to handle programming changes for a robot. These factors plus high (and often unanticipated) costs and risks combine to make the ROI hurdle rates for investment in AMES very high.

Most current U.S. robot users report a payback period of about two years. Industrial robots operated for three shifts (as in many machine load/unload applications) generally have shorter payback periods, as little as one year. Robots operated on only one shift tend to be more sophisticated and expensive (e.g., for many assembly and painting applications), resulting in payback periods of three to four years. It should be noted that robots have been most widely adopted in those applications where robot capabilities and costs were established, and where payback is shortest—the "easiest" applications.

Optimistic forecasts of increasing demand for robots may be off the mark, since demand will be for increasingly "difficult" applications over time. On the optimistic side, increasing robot capabilities and decreasing costs, coupled with improved user perception and evaluation of the benefits of robots will favor their introduction.

As noted, the most important element of most current evaluations of AMES payback has been the reduction in direct labor costs, which is unquestionably significant. Unimation and Cincinnati Milacron cite the hourly cost of operation for their programmable, servo-controlled robots (types A and B) as $5.00 and $6.00 respectively. Hourly cost of operation in this case includes financial costs (amortization of investment at 15% borrowing cost, taxes, insurance, etc.), maintenance and repair, and robot consumables (energy, compressed air, water, etc.). The hourly cost (including fringe benefits) of labor that could be replaced by a Unimate in the automotive industry, for example, is now in the range of $17-20 an hour.[20]

High wage rates for auto workers have been a factor in that industry's lead in adoption of industrial robots. Yet wages in other areas of manufacturing may not be high enough to justify investment in AMES solely on the basis of reduced labor cost. The U.S. currently has a huge labor surplus (industry as a whole substituted labor for capital equipment throughout much of the 1970's),[21] and the rate of wage inflation has slowed during the current recession. Under these circumstances, users are less likely (or able) to invest in robots in order to capture labor cost savings.[22]

Leasing of automated manufacturing equipment has been suggested as a method whereby users could avoid the high costs and risks of purchase. Robot leasing has not yet caught on in the U.S.; weak demand affects leasing as well as sales of AMES. The cost of robot leasing in the United States (when and if available) is $1,000 a month for a three-year period; by comparison, the cost of leasing a simple robot (used to tend an injection molding machine) in Japan from JAROL is about $170 per month.[23] The difference in leasing cost reflects the difference in capital cost to the leasing agency (already lower in Japan, but further enhanced by explicit government policy).

Corporate tax laws in the United States have been amended to favor capital investments in general—more rapid depreciation of equipment, investment credit deductions, and lower overall tax rates—but the net effect of these changes on AMES demand may be minimal because they are not focused on the technological upgrading implicit in AMES investments. The current tax structure seems to encourage merger and acquisition as much or more than productive investments in AMES.[24].

Industrial Management Attitudes and Practices

Certain characteristics and practices of wide segments of U.S. industrial management have undermined effective demand for AMES. Two key elements in U.S. corporate management are: 1) the short-term and financially-dominated point of view in senior and manufacturing management, and 2) emphasis on efficiency in operations at the expense of long-term manufacturing stategy and structure. Top level management too often has failed to recognize the strategic importance of AMES as a determinant of competitiveness (and in some cases, of survival) in international markets. Instead, a growing number of U.S. firms have progressively moved towards short-term measures and tactical expedients as the basis of maintaining separate earnings. During the 1950's many firms adopted scientific marketing methods and emphasized new product development, advertising, and marketing as the basic competitive tools. Market research methods developed during the 1950's became highly sophisticated during the 1960's. Modern financial management techniques developed during the 60's emphasized annual operating plans and highly accurate financial measurement and control systems, focusing on profit centers and quarterly returns.

As a consequence, management promotion and incentive systems punish mistakes and reward success quarter by quarter. This is partly the result of the demands by major sources of investment capital (large pension and investment funds) for steady quarterly profits and dividends. Another reason for the supremacy of short-term financial management is the steady replacement of owner-management (with technical or manufacturing backgrounds) by professional management that increasingly demands business school graduates with a high level of analytical and quantitative skills.

These skills are essential, but have the unfortunate by-product of focusing attention on the kinds of actions that are easy to measure because they have large costs and benefits in the short term. This leads to a high degree of risk aversion on the part of corporate managers, regarding major investments in innovative technology (such as AMES) whose greatest benefits are difficult to quantify and are felt in the long term. Past experience with underutilized equipment that often required substantial and unanticipated additional investment in support facilities and overhead contributes to this risk aversion. As U.S. management is currently structured, it is usually more rewarding for the individual manager to avoid innovation and let competitors accept the big risks of early AMES installations.[25]

One of the clearest trends today among potential U.S. users of AMES is that many are delaying purchases until the "bugs have been worked out" of the more sophisticated equipment (e.g., intelligent robots with sensors), which would presumably eliminate many manufacturing inefficiencies at once. Many

Japanese companies, in contrast, have been rapidly implementing some fairly basic existing robotic technologies. By accepting the initial risk involved in early robotization, these firms achieve a lowered cost curve that increases competitiveness, provides capital for investment in more sophisticated AMES and so forth in a self-reinforcing cycle. Such risk-taking is possible only where management can operate within a long-term, strategic perspective.

No less a figure than the President of Ford Motor Company, Donald Petersen, has acknowledged that the major responsibility for restoring the growth of U.S. productivity lies with business management, and that the primary means of solving the problem is to manage for the long run. [26] If this realization has come to Ford somewhat belatedly, they are now making efforts to catch up: Ford is the second largest user of industrial robots in the U.S.

One aspect of a short-term, financially-oriented management style that contributes directly to reduced effective demand for AMES is the emphasis on payback periods and financial return on investment (ROI). When combined with the high interest rates of recent years, these methods of investment evaluation lead to increasing difficulty for the manager in justifying a proposal to purchase new production equipment. The hurdle rates for ROI in some industries have climbed to 35-50%, with the result that investments in AMES which could ensure long-term competitiveness are simply not made.

The motivations of current and potential users of robots provide an indication of the pervasiveness of short-term thinking in evaluation of AMES investments. A primary motivation for introducing robotics is to reduce direct labor costs. [27]

Arguments about improved quality or increased production flexibility are regarded as too nebulous and do not carry much weight with financial controllers, who may insist that ROI calculations will not be favorable without a drastic decrease in labor costs. This emphasis is changing somewhat, toward a secondary emphasis on "improved product quality" by prospective user firms actively investigating the purchase of a robot.

Improvements in product quality through robotization only become evident in the longer term (measured in years), while reductions in direct labor cost brought by robots can be felt within months. Japanese industry, and some American analysts, [28] have long recognized that increasing product quality and concomitant customer satisfaction is the basic essential for maintaining and expanding market share in international competition. If a growing portion of U.S. industrial management is stressing "improved product quality" in evaluation of investments in AMES, this increasing sophistication on the part of users may presage a more rapid increase in effective demand for AMES.

Some users have cited a desire to "project an image of innovativeness" and to "keep up with the Japanese" as reasons for investing in industrial robots. Yet most experts agree that the new automated manufacturing technology will not, in and of itself, solve the problems of high cost, low quality, and long lead times.

Quick-fix and cosmetic solutions (such as the installation of a few token robots) to problems of competitiveness in manufacturing reflect the low position of manufacturing (or 'operations' to use the current term) management within the corporate hierarchy. Because of the supremacy of short-term, financially-oriented attitudes, manufacturing managers tend to concentrate on short-term productivity and efficient operation of in-place systems, perhaps seeking marginal improvements in costs, quality, inventories and delivery. Operations managers are not encouraged to seek alternatives, and thus lack insights of what is possible and of what other firms are doing.

A recent informal poll of jewelry manufacturers asked what new technologies they saw impacting on the jewelry manufacturing process. While some mentioned robotics, over half that did so felt it had no application whatsoever to their industry. [29] Yet robot soldering, welding and assembly (as well as CAD/CAM and computer-controlled laser cutting and engraving) have already been used in some jewelry manufacturing operations.

Progressive U.S. top management realizes that the revolutionary potential of AMES can be fully realized only if more emphasis is placed on manufacturing within corporate strategy and structure. [30] Manufacturing has become an integral part of the long-term competitive strategy of a small but growing number of U.S. firms, some of which are, or could be important actors in the field of AMES. General Electric, for example, has developed a course for manufacturing managers specifically designed to more fully integrate manufacturing operations into top management's strategic business plan. Course participants are encouraged to learn "what is possible" through in-depth visits to highly productive companies, including several in Japan. Participants then undertake some new program—directly spawned by the course experience—for realizing significant productivity improvement.

Encouragement of risk-taking and strategic viewpoints among both manufacturing and senior management at GE is a reflection of the view that improving productivity (the efficient use of all the firm's resources) is a key element of long-term business strategy. In this view, improved productivity—and not simply lower costs—is the desired end. Costs, quality, dependability and flexibility are viewed as non-conflicting subsets that are means to this end. Thus, the manufacturing manager is free to focus on long-term productivity improvements rather than agonizing over annual programs to cut variable manufacturing costs.

In this type of management climate, a proposal to invest in AMES to achieve long-term productivity improvements will be more favorably received. GE has in fact begun to adopt AMES fairly rapidly, and has industrial robots in use in at least eight different operating divisions. GE has even established a "rent-a-robot" program to furnish units to its plants on a try-out basis. GE is admittedly one of a small number of U.S. firms to emphasize manufacturing as an element of corporate strategy. The experience gained as a large *user* of AMES will certainly benefit GE in its effort to become a large AMES *producer*. Yet, effective demand for AMES is inhibited by inappropriate management as described above, and GE's efforts (as well as those of other producers) to commercialize AMES technology could suffer.

Labor's Receptivity to AMES

During the three decades following World War II, labor leaders were concerned primarily with pay increases, benefits and job security, rather than job enrichment, the work environment and the quality of work life (QWL). During that period, expansion in both the workforce and the economy as a whole mitigated against rapid adoption of automated manufacturing equipment. Industrial robots, first introduced in 1961, found little acceptance or widespread interest for the next fifteen years.

During the 1970's, labor's concern with QWL issues increased, and the Occupational Safety and Health Administration (OSHA) was created partly in response to these concerns. One result of this change in attitudes was increased acceptance of automation in jobs that are unsafe or undesirable for humans. These jobs were among the first to be automated because the cost of labor to fill them was relatively high and the payback period for the robot replacement relatively short. Labor has usually encouraged the automation of such jobs and this encouragement has been a significant element in the growth of automated manufacturing from the mid-1970's to date. This factor is decreasing in importance, however, as fewer and fewer jobs that are clearly undesirable remain to be automated.

Another factor that has increased acceptance of AMES among both white and blue collar workers has been the growing belief that automation is not only necessary, but inevitable in order to increase productivity, get out of the recession, and regain international competitiveness. Labor leaders have come to realize that job displacement due to the introduction of robots is likely to be less than would occur if we failed to match the efforts of overseas competitors in industrial automation.[31] But net losses are bound to continue to occur as long as the economy is unable to expand into compensatory growth areas. Short-run job displacement due to automation is occurring now and will continue for some time.

A study of automation in the U.S. automotive industry found that each robot installed today displaces about two workers a day. As robot capabilities increase and applications shift to manufacturing areas with 3-shift operations, the ratio of robots to workers displaced will increase, so that by 1990 each robot will displace more than three workers a day.[32] But there is little agreement on what the overall job displacement effects of AMES will be in the long run.[33]

Industrial robots will continue to displace direct labor in several job categories, notably assemblers, inspectors, production painters, welders, packagers and machine operators. Employment creation and skill upgrading through introduction of AMES is unlikely to compensate for job displacement in the long term. Over half of all the jobs that could be replaced by industrial robots are in the five major metalworking sectors and is highly concentrated in geographic terms.

Role of Labor Unions

Labor unions face a difficult task in cushioning the effects of automation on their membership at a time when employers are doing their utmost to reduce costs. International mobility of capital gives management a powerful lever to undermine wages and working conditions, and management has been explicit about using it in discussions with unions.[34] Automated manufacturing, which can itself be internationalized, provides another such lever.

Although some labor unions are losing members, many of the blue collar workers most likely to be displaced by industrial robots are highly unionized. Many white collar workers affected by computer aided design and engineering systems are also unionized.[35] Union leaders, for the most part, are aware of the link between AMES, productivity, and international competition. Many have adopted a fairly sophisticated approach to the tradeoffs between employment and automation in negotiations with management..[36]

Clearly there has been a much higher degree of cooperation among the business community, labor unions, and government in planning and implementing programs to cushion displacement effects in Europe and Japan. Europe benefits from a history of much closer relations between management and labor, and Japan from their paternalistic relationship.

The job displacement effect of robotics has been softened by the fact that many robots were introduced only as worker attrition permitted, and by their early use primarily in undesirable jobs. These factors do not obtain in the long run, and American unions have emphasized retraining programs and other measures in recent contract negotiations. In the United States, these private sector measures have lagged far behind those in other countries, and public policies for labor market adjustment have not been nearly as effective as in Europe or Japan. Under these circumstances, the labor resistance (both union and non-union) to the introduction of automation has been stronger here than abroad. As high unemployment and the recession continue, its negative effect on demand for AMES has tended to increase as short-term job displacement has grown.

FOOTNOTES

[1]See Chapter VI for a comparative overview of the United States relative to Japan and Western Europe.

[2]See "U.S.-Japan Trade: Issues and Problems," report by the U.S. Comptroller General, ID-79-53 (21 September 1979), pp. 164-65.

[3]Robot Institute of America, "The Decline of Productivity and The Resultant Loss of U.S. World Economic and Political Leadership," Policy Document No. 2 (March 1981), pp. 2-3.

[4]D. Wisnosky, "The Factory of the Future," testimony before the Labor Standards Subcommittee, U.S. House of Representatives (23 June 1982), p. 8.

[5]R. Kirkland, "Taxing the Business Lobby's Loyalty," *Fortune*, 18 October 1982, p. 144.

[6]A recent study of 13 government funded R&D programs on manufacturing technology (in the areas of automation, machine and tool design, and materials handling) found that in terms of dollars and scope of activity, the programs were dominated by military applications and emphasized the solution of particular manufacturing problems through development of single machine technologies. See Hetzner, W., et al., "Manufacturing Technology in the 1980's: A Survey of Federal Programs and Practices," National Science Foundation (3 September 1981), p. 12.

[7]See full page ad in *Washington Post* (11 June 1982), p. A8.

[8]See, e.g., J. Albus, "The Federal Role in Robotics: The Fledgeling Stage," in *Enterprise* (National Association of Manufacturers, August 1982).

[9]For a detailed description of military involvement in AMES, see E. Martin, "Department of Defense Statement on Robotics Technology," Subcommittee on Investigations and Oversight of the Committee on Science and Technology, U.S. House of Representatives (23 June 1982).

[10]For a description of early developmental patents leading to industrial robots, see R. Ayres, et al., *Technology Transfer in Robotics Between the U.S. and Japan* (Pittsburgh: Carnegie-Mellon University, 1981).

[11]L. Conigliaro, *Robotics Newsletter*, No. 9, 14 June 1982 (New York: Bache, Halsey, Stuart, Shields, Inc.).

[12]See Part III for profiles of leading U.S. companies.

[13]For GE's viewpoint, see D. Kinney, "Manufacturing Technology - Part of the Solution." *Proceedings of the 13th Annual DOD Manufacturing Technology Conference— MTAG 81* (30 November-3 December 1981), pp. 305-313.

[14]Joseph Engleberger of Unimation has said, "We never really wanted to do the work of integrating our robots with the other machines in a customer's factory." Quoted in J. Dizard, "Giant Footsteps at Unimation's Back," *Fortune*, 17 May 1982, p. 98.

[15]By way of contrast, see profiles (Part III) on Dainichi Kiko (Japan) and GCA (United States), its joint venture partner.

[16]H. Tuttle, "Surge to Robots On," *Production,* December 1981, p. 104.

[17]R. Ayres and S. Miller, "Industrial Robots on the Line," *Technology Review*, May/June 1982, p. 38.

[18]R. Sanderson, et al., *Industrial Robots* (Naperville, Illinois: Tech Tran Corporation, 1982) p. 84.

[19]*Industrial Robots, ibid.*, p. 96.

[20]See G. Munson, "Commentary," before Subcommittee on Labor Standards, U.S. House of Representatives, 23 June 1982, p. 9.

[21]Capital investment per worker has fallen in the United States at the same time it has increased in competitor countries. See, e.g., Robot Institute of America, "The Decline of Productivity and Resultant Loss of U.S. World Economic and Political Leadership," Policy Document No. 2 (March 1981), p. 3.

[22]See Table 2, above, for a comparison of robot-worker densities in the United States, Japan, and Western Europe.

[23]C. Waters, "There's a Robot in Your Future," *INC.* magazine (June 1982), p. 68.

[24]The recent merger activities among Bendix, Martin-Marietta, United Technologies Corporation and Allied Corporation provide a prime example. Cash-rich Bendix (an AMES producer and user) tried to acquire Martin-Marietta (a high-technology firm that could benefit from greater investment in AMES), who in turn invited United Technologies (another AMES user rumored to be interested in becoming a producer) to join in the Bendix take-over. Eventually Allied Corporation (an oil and chemicals firm) bought into both Martin-Marietta and Bendix. See R. Rowan and T. Moore, "Behind the Lines in the Bendix War," *Fortune*, 18 October 1982, p. 156.

[25]For the mirror image of management perspectives, see Chapter V below for discussion of Industrial Management and Labor Relations (Japan).

[26]Donald Peterson, "Answer to America's Problems—Managing for the Long Run," *Production,* September 1981, p. 35.

[27]See R. Ayres and S. Miller, "Industrial Robots on the Line," *Technology Review*, May/June 1982, p. 43.

[28]See W. Abernathy, et al., "The New Industrial Competition," *Harvard Business Review,* September-October 1981.

[29]S. Aletti, "A Revolution Coming," *American Jewelry Manufacturer*, September 1982, p. "D".

[30]Wickham Skinner, *Manufacturing in the Corporate Strategy* (New York, New York: John Wiley & Sons, 1978).

[31]See R. K. Vedder, "Robotics and the Economy," staff study prepared for the Subcommittee on Monetary and Fiscal Policy of the Joint Economic Committee, U.S. Congress, 26 March 1982.

[32]W. Tanner and W. Adolfson, "Robotics Use in Motor Vehicle Manufacture," U.S. National Technical Information Service, February 1982, p. 107. The article points out that: ". . .more than two thousand direct labor positions have already been filled by robots in North American automobile factories. These robots have, however, created about 250 new positions in the skilled trade ranks. By 1985, under the most modest growth rate anticipated, another 10,000 direct labor jobs will be gone, while 1,000 new skilled trades jobs will be added. If the introduction of robots is pursued aggressively, the loss of direct labor jobs by 1985 could approach 30,000 with fewer than 3,000 new positions generated."

[33]For a discussion of this point, see B. Usilaner, "Automation in the Workplace," Statement before the Subcommittee on Labor Standards, Committee on Education and Labor, U.S. House of Representatives, 23 June 1982.

[34]See, e.g., W. Serrin, "The Mobility of Capital Disperses Unions' Power," *New York Times*, 21 March 1982.

[35]See D. Chamot, "Testimony," before the Subcommittee on Labor Standards of the Committee on Education and Labor, U.S. House of Representatives, 23 June 1982.

[36]For example, companies are responsible for retraining displaced workers, relocation allowances are provided, advanced notice must be given on proposed technological changes, and in some instances there are arrangements to share in the profits to be realized from introduction of AM technologies.

CHAPTER V
JAPAN

National Environment

Government Industrial Policy [1]

The government-industry relationship in Japan plays a major role in the process of formulation of Japan's economic and technological development policies, as well as in their implementation. The Japanese government has strong confidence in competition and the market forces, but it also perceives a need to guide market forces from time to time in order to achieve and maintain a high level of performance of the economy. There has, for example, been an unrelenting effort to encourage the movement of people and resources into sectors with high growth and high productivity. Areas of improving comparative advantage are encouraged to accelerate, and declining or poor-performance industries are encouraged to phase down.

The Japanese government is now emphasizing the so-called knowledge-intensive industries, which include robotics. The "vision" for the 1980's calls for steadfast movement toward a "technology-based nation." The strong commitment to knowledge-intensive industries can be traced back to the early 1970's, when Japanese government studies, prepared in cooperation with Japanese industry, predicted the development of information-oriented societies and set out agendas to develop the complementary knowledge-intensive industries.

Each of these areas represents progressive requirements in precision, quality control, and cost effective manufacturing techniques to incorporate the new demands. It is part of the overall efforts in miniaturization of product components, at which the Japanese have been commercially successful in successive product areas, beginning with items such as cameras and related optical goods and continuing on to videotape recorders and related consumer electronic goods, and more recently into the office and home computer field. [2]

Formal Structures

Besides the major role of the Ministry of International Trade and Industry (MITI), significant economic policy-making role is also played by the Economic Planning Agency (EPA) of the prime minister's office, which does long-range planning for economic growth and suggests areas where government support and action can best advance these plans. These long-range plans are laid out in close consultation with academicians, consumer groups, and industry (the last through advisory committees composed of business leaders and through business groups). The ministries and various independent agencies formulate their own plans, but take into account the EPA ideas.

Responsibility for implementing policy guidelines resides principally in two ministries: The Ministry of Finance (MOF) and MITI. The MOF is the ultimate source of financing, and by guiding commercial banks, it influences the direction of financing of industrial investment in Japan and has intimate links with the entire banking system. To maintain and expand Japan's industrial growth while minimizing commercial risks, the banks and other investors tend to rely on the MOF-MITI perceptions of what are the most promising avenues of expansion and innovation.

The Technology Component in Government-Industry Relationships

A wide array of Japanese government policies and consultative procedures are designed to further strengthen the country's already formidable industrial technology base. The government, in consultation and general agreement with broad sections of Japanese industry, has established several guidelines and objectives in its efforts to stimulate further the country's technological development. Emphasis is placed on the commercial application of current state-of-the-art capabilities and, increasingly, on the development and commercialization of next-generation technologies.[3]

Cooperative Agreements

The Japanese government uses multiple cooperation contracts with the largest cooperations to develop technical innovation, especially if it promises to contribute to the enhancement of international competitiveness. This occurred in the automotive industry during the 1950's and early 1960's, and in the nascent computer industry during the 1970's. The 1978 Temporary Law for the Promotion of Specific Electronics and Machinery Industries promotes the aggregation of the industries by allowing cartel formation and allowing the government the right to oversee the plans of companies operating under the law.

The Japanese government encourages and facilitates cooperation between firms in the same industry, thus avoiding the antitrust restrictions of the United States. Companies share information on new products under development in order to eliminate repetition of costly research and engineering. Since so much of Japanese industry is geared for export, this acceptance of industrial cooperation strengthens Japan's competitiveness in the world markets.

Tax and Depreciation Measures

Japanese fiscal policy has a number of provisions designed to enhance the nation's technology positions. If a Japanese firm's R&D expenditures for a given year exceed the largest amount of annual R&D expenditures for any preceeding year since 1966, 70 percent of the excess may be taken as a credit against the corporate income tax. Firms which are members of research associations can take an immediate 100 percent depreciation deduction on all fixed assets used in connection with research activities.

Government Funded R&D

While most R&D work in Japan is performed and financed by industry, government laboratories do conduct a limited, but significant, amount of R&D work in specific, well-targeted areas. These government labs are usually involved either in basic research or new product conception. This targeted R&D activity by the government has played a substantial role in the dynamic growth of several Japanese industries, such as electronics and automotive in previous years, and currently in new high technology areas such as robotics and biogenetics. In performing these activities, the Japanese government is, in effect, reducing the risk and cost to Japanese industry of developing new technologies.

The Japanese government is also involved in providing direct financial assistance to the private sector to develop its technological capabilities. The various types of government financial assistance all represent a form of public subsidization of private technology development efforts. While the absolute amount of subsidization has not been large in recent years, it is well-targeted to the promotion of specific industries and products which hold the potential for attaining world technological leadership.

Japan spends very little money on defense-related R&D, allowing it to concentrate its funding projects with greater potential for private sector commercialization. But in the future, there will be limits to the Japanese government's ability and willingness to fund R&D. The 1982 budget deficit is approximately $40 billion; and the Ministry of Finance wants to cut back on funding research even in areas everyone has agreed to promote.

It should be stressed here that Japan's fierce competitiveness is based not only upon its ability to research and develop new technology, but also upon its proven proficiency in accumulating and improving upon technology developed elsewhere.

Government Measures Regarding Robotics

The government uses a variety of financial and other incentives to encourage companies to enter growing industries, such as robotics. It is felt that domestic competition will lead to success in the international market, where much of Japan's rapid economic growth has come from. The government is presently offering support to AMES producers to facilitate the proliferation of companies in the market. The government's promotional role has been aimed largely at intensifying competition and reinforcing demand for AMES. In fiscal year 1980, the following measures were implemented: a) A leasing scheme by the development bank of Japan; b) Computer controlled industrial robots were added to the list of equipment eligible for special depreciation under corporate tax liability; and c) Special financing for robots was provided by the National Finance and Small Business Finance Corporation for robots with industrial safety features and for purchases of robotics to modernize small business operations.

Japan's Ministry of Trade and Industry (MITI) considers the robot industry as strategic for Japan's future for several reasons.

First, robotics is an industry which is expected to maintain a high level of growth, domestically and internationally.

Secondly, robotics increases productivity mainly by reducing costs and reallocating labor to more productive usage.

Thirdly, desired restructuring of the Japanese economy is facilitated through government support of robotics. For example, support of robotics fosters "knowledge intensification" in other sectors of the economy, and contributes to energy conservation and diversification.

Fourthly, in an expanding economy robotics contributes to job stability and guaranteed employment, in accordance with government labor policies. The worker is thereby insulated from extreme fluctuations in the machine tool industry and work is more humanized, less hazardous, and more productive through retraining for more highly skilled jobs.

Finally, robotics aids in government efforts to explore and develop the resources of the ocean and outer space.

Active intervention by the Japanese Government in support of robotics began in 1978 when the industrial robot was officially designated as an "experimental research promotion product" and as a "rationalization promotion product," under the special Law on Extraordinary Measures for Promotion of Specific Machinery and Information Industries. Having identified robotics as a major strategic industry for Japan's economy, MITI implemented specific measures to encourage research, development, and commercialization of robotics.

1. *Financial Incentives.* Financial incentives are a key element in the Japanese Government's efforts to promote robotics. Low-interest, direct government loans are available to small- and medium-scale manufacturers to facilitate the installation of robots. In fiscal year 1980, approximately $26 million (5.8 billion yen) was budgeted by the Japanese Government for these loans through a government agency, the Small Business Finance Corporation. In addition to low-interest direct government loans, a supplemental depreciation of 13 percent above normal depreciation is available to companies utilizing robots. An industrial firm can, by installing an industrial robot, depreciate 52.5 percent in the first year.

2. *Robot Leasing.* MITI was also influential in the founding of a robot leasing company, *Japan Robot Lease* (JAROL), in April 1980. The major shareholders are the 38 leading industrial robot manufacturers, three robot consultants, and 17 marine insurance companies. The principal shareholders have a capitalization of 300 billion yen. The purpose of JAROL is to facilitate the leasing of robots at low rates through leasing, selling, installment payments, consulting, and the supplying of information about the industrial robot and its application to industry. Sixty percent of the operating funds are financed by low-cost loans from the government's Japan Development Bank, the Long-Term Credit Bank, the Industrial Bank of Japan, and certain commercial banks.

The consortium uses this capital to purchase robots which are then leased primarily to smaller firms. The program is similar to that operated by the Japan Electronic Computer Corporation, organized more than 20 years ago under MITI auspices to support Japanese-owned computer firms. In addition to leasing robots, JAROL's staff provides engineering assistance for installation and programming.

In its first year of operation, JAROL's 52 leasing contracts amounted to 1.15 billion yen (around $5.5 million). JAROL is now offering flexible two to three year rental agreements and also plans to facilitate overseas leasing of robots through loans by the Japan Export and Import Bank.

In 1981, robot production was valued at 120 billion yen: 36 billion was consumed within the company and 84 billion outside, of which 8.4 billion was leased. JAROL itself leased 4 billion yen worth of robots, mostly to small companies.

The average value of leased robots has dropped from 10 to 12 million yen in 1980 to between 7 and 8 million yen in 1982.

JAROL was originally the leader in the field. Today there are about one hundred leasing companies operating in Japan.

3. *Research and Development.* MITI initiated a seven-year, 30 billion yen (around $150 million) robot research program in April 1982. This program will be organized much like other cooperative R&D programs in Japan, such as the well-known program aimed at very large scale integrated circuits (VLSI). About ten commercial companies are expected to be involved, plus the Electrotechnical Laboratory of the Agency for Industrial Science and Technology.[4]

The robot program is in part a sequel to previous MITI sponsored work on remote control devices for manufacturing and repairing the radioactive portions of nuclear power plants. The current robot program, however, will be much broader.

AMES Producers [5]

Overview of Industry Structure

The robotics industry in Japan has followed a familiar pattern of development established by many other industries, such as autos and computers. The process begins with the importation of foreign technology and equipment. Japanese firms then improve upon the new technology and begin to produce and distribute their own models. These new models filter down through the Japanese market, beginning with the larger companies and then working their way down to the smaller firms that need to absorb the new technology in order to compete. As the domestic market becomes saturated, Japanese producers focus their attention on exports in order to increase production and expand their market.

In 1967, Japan imported its first Versatron robot from the United States. The following year, Kawasaki Heavy Industries signed a technology license agreement with Unimation and began production. The robotics industry grew slowly in its first few years because of the relatively cheap labor available in Japan. But following the oil shock of 1973, the demand for robots picked up considerably. Wages increased to keep pace with the rising prices of consumer goods, while at the same time the price of a robot actually decreased. Between 1976 and 1980, robot production quadrupled, increasing from 4,400 units to 19,900. In value terms, the increase was ten-fold, from 10 billion yen to 100 billion yen, (approximately US $50-500 million), reflecting the spread of the more expensive sophisticated robots. By 1980, Japanese production of robots had expanded to more than fill domestic needs. In that year only 2 units were imported, whereas 1,170 were exported, accounting for a mere 2 percent of total sales.

There are roughly 150 Japanese companies producing robots.[6] Market shares for all producers are small. Because of the large number of producers and also because of the diversity of user demand, small companies which specialize in only one type of robot are able to survive in the market. Very few companies produce robots only; indeed, robot sales of even the industry leaders account for only a small percentage of that company's total sales.

Many Japanese companies produced robots initially to meet their own production needs. They did their own research, development and engineering and were able to test and improve their robots before offering them on a commercial basis. Consequently, Japan developed not single units, but entire manufacturing systems of which robots were an important part, but only a part. Japanese companies began by developing the simplest manipulator robots, then gradually turned to more sophisticated robots. These sophisticated, more "intelligent" robots are now beginning to predominate the market, and are expected to further increase their share of the product mix in the years ahead.

Although exports accounted for only 2 percent of total sales in 1980 and a little more than 3 percent in 1981, it is expected that exports will grow substantially in the years ahead. Many estimates project that exports will claim as much as 20 percent of production by 1990. Since overseas direct sales will require an extensive service and maintenance network, Japanese companies will most likely opt for licensed production and joint ventures in the near term. This will also help reduce the barriers presented by the growing trade friction directed against Japan. In time, Japanese firms may develop their own sales and distributing networks overseas.

Corporate Strategies and Outreach

There is no single model of corporate strategy that adequately provides a consensus view of Japanese companies. For one thing, levels of R&D spending and the attention paid to marketing have increased during the last few years. However, the manufacturing process itself has remained an integral part of virtually all companies' corporate strategy.

Japan today spends a greater percentage of GNP on R&D than the United States. This is a recent development and is due primarily to increased industry spending on R&D. The Japanese government still spends far less on R&D than the United States, even if defense-related R&D is discounted. This increase in the level of R&D spending manifests itself especially in the robotics industry, where Japan has chosen to take the lead instead of merely improving upon foreign technology.

Fujitsu Fanuc for example, has been successful in the robotics industry because of its strong orientation towards technological growth and

TABLE 7

JAPANESE PRODUCERS OF AMES

Name of Company	Value of Robotic Equipment Sales 1981-1982 (yen million)*		Robotic Equipment Sales as % of Total Sales		Main Product Fields	Type of Robotics Equipment	User Industries
Aida Engineering Ltd.	1981	1,000	3.2		press machines, automation devices	pressing	electric appliance, automotive
	1982	1,400	4.5				
Amada					band-saw lathes, press machines	computer assisted manufacturing systems	machine tool makers
Dainichi Kiko Co., Ltd.	1981	1,160	100		industrial robots	multiple uses (arc welding, assembly handling, palletizing)	
	1982	2,200	100				
Fujitsu—FANUC Co., Ltd.	1981	800	1.0		NC devices, NC machine tools, robots	cutting and grinding process, memory and processing	machine tool makers
	1982	2,000	2.3				
Hitachi, Ltd.	1981	1,550	0.1		heavy electric machines, household, electric appliances, communication electronic equipment, industrial machines, transportation machines	arc welding, multiple uses, painting	automotive
	1982	2,200	0.1				
Kawasaki Heavy Industries, Ltd.	1981	4,730	0.7		ships steel structures motors, machines	painting, arc welding, assembly, spot welding	automotive
	1982	7,000	0.9				

*Average: 250 yen = US $1.00

Name of Company	Value of Robotic Equipment Sales 1981-1982 (yen million)*		Robotic Equipment Sales as % of Total Sales	Main Product Fields	Type of Robotics Equipment	User Industries
Kobe Steel, Ltd.	1981	960	0.1	iron and steel products, machines, non-ferrous metals	painting, arc and spot welding, base alloys	automotive, plastics, metalic furniture
	1982	2,000	0.2			
Komatsu				forklifts, integrated distribution, machinery maker	arc welding	within company
Matsushita Industrial Equipment Co.	1981	–	–	distribution equipment, electronic equipment, condensor welding machine	arc welding	
	1982	600	1.27			
Mitsubishi Electric Corporation	1981	–	–	heavy standard electrical equipment, household electrical appliances, electronics and industrial equipment		machinery, construction machinery
	1982	1,100	0.19			
Mitsubishi Heavy Industries, Ltd.	1981	1,450	0.1	ships and steel structures, prime movers, chemical plants, construction, precision machines, engines, heating and cooling machines, aircraft, special vehicles	painting, spot welding inspection and measurement	automotive, home appliance makers and related type of industry
	1982	2,000	0.1			
Murata	1981			ceramic capacitors, resistors, piezoelectric products, circuit components		

*Average: 250 yen = US $1.00

TABLE 7–Continued

Name of Company	Value of Robotic Equipment Sales 1981-1982 (yen million)*		Robotic Equipment Sales as % of Total Sales	Main Product Fields	Type of Robotics Equipment	User Industries
Nochi-Fujikoshi Co.	1980	500	0.5	bearings, tools, hydraulic machines, machine tools, materials of steel	painting, spot welding, arc welding	
	1981	1,070	1.1			
Nippon	1981	220	100	robots	die-casting	automotive
	1982	230	100			
Osaka Transformer Co., Ltd.	1981	1,080	2.6	electric machines, electric welding materials	arc welding	
	1982	1,800	4.5			
Pentel	1981	–	–	drawing and writing materials, electronic equipment, robots for assembly	assembly	electronic, electric industry, automotive, precision machine, industry
	1982	250	0.8			
Sankyo Seiko Mfg. Co., Ltd.	1981	–	–	music boxes, watch machines, electronic parts, information machines, movie cameras, machine tools	assembly	
	1982	700	1.3			
Shinmeiwa Industry Co., Ltd.	1981	1,250	1.7	machines and equipment for aircraft and specially fitted vehicles	arc welding, cutting (plasma), shearing robots	construction, machine field, heavy industry field, agricultural
	1982	2,400	3.0			
Tokiko Ltd.	1981	500	0.7	automobile equipment, hydraulic machines, pneumatic pressures equipment for instrumentation	painting	home appliances, cameras, office machines, automotive, kitchen appliances
	1982	1,000	1.4			

*Average: 250 yen = US $1.00

Name of Company	Value of Robotic Equipment Sales 1981-1982 (yen million)*		Robotic Equipment Sales as % of Total Sales	Main Product Fields	Type of Robotics Equipment	User Industries
Toshiba Seiki Co., Ltd.	1981 1982	800 1,000	17.0 (1980)	general electric machinery maker: home electric appliances, communications, electronic machinery	assembly	
Yaskawa Co., Ltd.	1981 1982	2,700 4,500	31.8 45.0	power transmission gears, power material handling machines, robot material handling machines, information processing equipment	arc welding, material handling, painting	automotive construction, electricity, agricultural appliances

SOURCE: Data sheets furnished by Japan Industrial Robot Association.

*Average: 250 yen = US $1.00

development. In 1979, some 200 of its 730 employees were in R&D, compared to 330 in sales and service and 200 in production. This is an extreme example of a company willing to devote its skilled engineers and resources to R&D. Komatsu has also centered its corporate strategy on research and development, mainly for the purpose of developing intelligent robots. Other companies, however, still rely on outside technology for production needs. Kobe Steel has emphasized borrowed technology to make various improvements in its robot models. Osaka Transformer Company, Japan's second largest producer of arc-welding robots is currently being charged with patent infringement by U.S. firms who claim that Osaka technology was copied from their models.

The government frequently sponsors joint R&D projects among several firms, research institutes, and universities. In light of the lifetime employment system, in effect with larger companies, this is particularly useful because it facilitates the flow of technological information between companies that is otherwise blocked. Some managers and engineers do not move from one company to another, nor does their know-how. Joint projects prevent costly duplications.

Japanese companies have always emphasized manufacturing strategy. This is precisely the key argument for introducing AMES: it cuts production costs and improves quality control. The question for Japanese companies is no longer whether to automate or not, but rather one of how much capital investment is justified by the volume of production. Capital investment can be justified when the business cycle is positive, when large volumes are produced, and high growth is expected. Capital investment in AMES is a sound alternative for those companies needing to move along the learning curve to achieve cost efficiency. Indeed, for some companies it may be the only alternative. Many Japanese firms, including Kubota Ltd., Fanuc, and KHI, run only single-shift days because of the reluctance of workers to work other shifts. Firms are unwilling, and in part unable, to hire workers to man a new shift. The only alternative to allow increased production is to automate and run multiple shifts. This round-the-clock production also provides a faster payout for the investment of capital.

Marketing has acquired new strategic importance for Japanese companies. Because so many firms in the past produced robots primarily for inhouse needs, the building of sales networks was unnecessary. But with the increasing commercialization of AMES, and particularly with the emphasis on exports, this has changed.

Robot producers frequently set up their own marketing division or link up with a trading house. Mitsubishi Heavy Industries' strategy is to link up with foreign manufacturers for co-production and marketing law royalties and to sell certain criticial components. Yamazaki, which exports about 70% of its exports, provides extensive training for its foreign customers who are more demanding than Japanese customers for applicator engineering and after-sales service. The strong overseas orientation of Aida is exemplified in its recently concluded contract with West Germany's Estel Hoesch Werke AG. Aida will provide technological aid involving CAD/CAM systems of metal molds. This is the first time that a Japanese company has exported to Europe its CAD/CAM system.

Forward Outreach to Users

As explained elsewhere in this paper, the markets targeted by Japanese robot producers are those industries which perform the most repetitious or the most hazardous tasks. At present, the largest user of robots by far is the auto industry. Automobile manufacturers, and its related industries, use 90% of all spot-welding robots. These are relatively unsophisticated robots and their importance to the industry will surely decline as more intelligent robots are developed and put into operation. Another example of an unsophisticated robot in wide use is the "pick and place" model used, for instance, in the assembly of ball point pens.

Since many of Japan's robot producers began producing them to meet their own needs, it is now surprising that they cover a wide variety of specific uses. Such companies as Yamaha and Honda developed robots for inhouse needs and did not market them. Other companies, such as Sailor Pen, Seiko and Pentel developed robots for their own assembly purposes and then marketed them for related functions. Machine makers that developed their own robots such as Fujitsu Fanuc and Toshiba, were able to market a whole manufacturing system instead of a single robot. This type of overall production system, incorporating not only assembly robots, but also automated transport and loading and unloading equipment, computer aided design and manufacturing, and centralized coordination is a promising future trend for robot producers.

Companies which originally developed robots for their own needs are obviously at an advantage in supplying these integrated manufacturing systems. Years were spent perfecting these robots and related equipment before they were offered to the outside market. Today, with the rapid development of intelligent robots, these flexible manufacturing systems (FMS) are becoming more practical and as a consequence more available.

The ultimate goal is to have robots producing robots with no human input. With the opening of Fujitsu Fanuc's experimental robot assembly plant in December 1980, this goal came one step closer to being realized. Other fully automated plants that produce goods ranging from instant noodles to machine tools are already in operation.

The Japanese market is by no means saturated, but a key strategy for the rapid development of the industry as devised by the government and industry leaders is a boldly concerted drive to market Japanese robots overseas. This marketing strategy coincides with the development of intelligent robots. U.S. firms in particular have shown greater interest in the expensive, sophisticated robots and have virtually ignored the more rudimentary assembly robots. Japan has taken the lead in developing intelligent robots and is counting on finding a lucrative market waiting for them.

Japan has been counting on large increases in robot exports to expand their production. But American and European resentment of Japanese trading practices will likely inhibit fast growth in exports, especially if foreign countries perceive Japanese robots are having a directly adverse effect on domestic unemployment. The current "Buy American" hysteria is a clear example of this. This protectionist attitude may prevent the importation of Japanese robots if it is felt that the robots will take away U.S. jobs. And with the prolonged recession still active, American industry is in no position to afford the costly investments needed for reindustrialization that robots will surely be a part of.

Several of the major manufacturers of robots, e.g., Hitachi and Toshiba, have been using robots within their own companies. With this arrangement, there are no gaps between the producer and user because they are the same firm. Other firms sell their robots to sister firms and have developed specialized robots accordingly. Mitsubishi Heavy Industries and Mitsubishi Electric developed robots to be used in Mitsubishi Motors. Mitsubishi Electric also produces a window washing robot to be used by Mitsubishi Electric, the owner of many of the tall buildings in Tokyo's business district. Toyota Machine Works began producing robots to be used by the auto manufacturer Toyota.

Other firms have hired trading houses to handle the marketing arrangements or have established their own. Robots are not sold as individual items but as part of an integrated manufacturing system. Therefore, the initial sale to a user is the most important because so much correlated production system knowhow is involved. Kawasaki Heavy Industries, which holds a near monopoly on the spotwelding robot market, switched trading companies because its former agent lost the lucrative Toyota contract. The larger producers contract with several trading companies to reach specific markets.

Backward Linkages to Component Suppliers

As stated in the previous section, many of Japan's robot manufacturing firms began building robots for their production needs. This not only eliminated the gap between producer and user but also between producer and supplier. Firms designed their own parts and equipment to suit their own intrinsic needs. When these firms began to market their robots, the inhouse supply network was already functioning smoothly.

Firms that are unable to produce all their own components needed for robot assembly rely upon a network of small supplier firms. Japan has a highly developed supply network with a number of small companies working in close cooperation with the industrial producers. To ensure quality control, contracting firms are given the privilege of making frequent on-site inspections. The small suppliers are continually encouraged to find ways of cutting their production costs without compromising the high quality of their work. In fact, to make these supplier firms more cost efficient, some robot producers are encouraging their suppliers to automate their manufacturing processes, bringing the supplier-producer-user relationship full circle.

The effectiveness of Japanese firms depends heavily upon the support of component and part subcontractors. Many individual subcontractors rely upon a single firm for a large percentage of their annual sales. The design and production capabilities of equipment manufacturers are generally linked to an intricate network of subcontractors whose quality and prices they are able stringently to control. It is for this reason that Japanese manufacturers are seriously inhibited when they attempt to set up overseas manufacturing operations, since they generally encounter considerable difficulties in duplicating the intricate networks of reliable subcontractors that can deliver quality components at low costs.

Product Spread and Market Share

It is difficult to state authoritatively the extent of Japan's dominance of the robot industry because there is not a universally agreed upon definition of what a robot is. The Japanese classify their robots into six categories:

1. Manual manipulator: directly operated by an operator.

2. Fixed sequence robot: a manipulator whose work process has previously been programmed. This programming cannot easily be changed.

3. Variable sequence robot: a manipulator whose programming can easily be changed.

4. Playback robot: a manipulator that has memorized a work sequence taught to it by a human operator. This work sequence is then repeated automatically from memory.

5. Numerical control robot: manipulator whose work is determined by numerical data regarding work sequence, position, and conditions.

6. Intelligent robot: uses sensors to determine position, action, and rate of production. Can detect changes in the work environment and uses its decision-making capacity to adjust accordingly.

By its own definition, Japan has produced and installed more than 75,000 robots. If manual manipulators and fixed sequence robots are excluded, the number falls to 14,000. This more restricted definition is the one used by the United States, which has roughly 4,500 installed, operating robots. Even with this more limited definition, Japan's edge is unarguable.

Japan's lead in the utilization of robots is likely to expand in the future. Japanese industry began by installing the simpler robots in large numbers and then gradually turned to more sophisticated units. As the industry moves toward these sophisticated robots, there is a greater potential for mass production and higher sales. The production of playback and intelligent robots has already increased to over one-fourth of the market. Because Japan has moved so far along the learning curve, its continued dominance of the industry seems assured.

The Japanese Industrial Robot Association (JIRA) estimates that the broader robot market will exceed $1.5 billion by 1985 and $2.6 billion by 1991. When compared to the 1981 sales figure for new robots of $350 million, JIRA's estimate shows the rapid growth expected in the robot industry. And as the trend towards the more sophisticated, and more expensive, robots continues, the sales volume and value will increase accordingly. The market share of these robots, as defined by U.S. terms, should rise to 80% or more. Other unofficial estimates are more optimistic than those put forth by the conservative JIRA. Some expect annual sales of sophisticated robots to rise to $2.2 billion by the mid-1980's and twice that amount by the mid-1990's.

The robot industry in Japan is highly competitive and diverse. As of 1981, there were at least 150 firms producing robots. The size of these firms varies considerably, as can be seen in Table 8. Market shares are small for all firms. Kawasaki Heavy Industries (KHI) sells 80% of the spot welders sold in Japan, has the largest market share of 8%. The two largest U.S. robot manufacturers, on the other hand, control 70% of the domestic market. Moreover, most Japanese producers do not depend on their robot sales. KHI's robot production accounts for only 1% of its over-all sales.

The diversity of the market allows for greater specialization. Most robots need to be custom built to fit the user's needs. This allows even the smallest firms to survive in the market by specializing in one type of robot. The greater flexibility provided by intelligent robots may allow large firms with mass production capabilities to eventually squeeze the smaller firms out of the market.

TABLE 8

SIZE OF ROBOT-PRODUCING COMPANIES IN JAPAN

BY SIZE OF CAPITAL

Capitalization	Number of Firms	%
Less than 10 million yen	19	14.3
10 - 100 million yen	36	27.1
100 million to 1 billion yen	23	17.3
1 - 3 billion yen	8	6.0
More than 3 billion yen	47	35.3
TOTAL	133	100.0

BY NUMBER OF EMPLOYEES

Number of Employees	Number of Firms	%
Less than 50	33	24.8
50 -100	29	21.8
500 - 1000	15	11.3
1000 - 5000	25	18.8
More than 5000	31	23.3
TOTAL	133	100.0

SOURCE: "Robots Revisited: One Year Later," Paul Aron Report No. 25, pp. 12-13. Daiwa Securities America, Inc., New York, New York, 1981. Permission granted to reprint.

Japanese users often request modification of standard designs to fit their individual needs. This is particularly problematic in responding to the needs and demands of the broad array of small-to-medium size firms in Japan. For these markets, Japanese AMES manufacturers maintain substantial industrial extension services for product process engineering and maintenance/service support.

Joint Ventures, Licensing Agreements, and Acquisitions

When Japan began its robot production more than a decade ago, it signed licensing agreements with U.S. companies. Japan has gone from being a net importer of robotics technology to become the world's leading producer and supplier of robots. Today, foreign companies seek out their Japanese counterparts to enter into joint ventures or licensing agreements.

Japanese companies will likely expand these tie-ups with foreign countries in order to facilitate its exports of robots. These tie-ups will allow Japan to enter foreign markets without the need to develop marketing and service networks. It should also diminish foreign resentment towards Japanese infiltration and dominance of yet another high-technology industry.

The original Japanese-U.S. licensing agreement between Unimation and KHI has seen a reversal of the flow of information provided by the relationship. Today, KHI offers its experience and expertise to its former benefactor. Other U.S. companies who have recently committed themselves to automated production have linked up with Japanese robot manufacturers. Examples of this are the Original Equipment Manufacturers (O.E.M.) agreements between Westinghouse and Mitsubishi Electric and Komatsu Ltd. Such tie-ups are tacit proof that Japan has gained the edge in the commercialization of robots.

One of the most significant cooperative arrangements is the joint venture of General Motors and Fujitsu Fanuc, known as GMF. It will be an autonomous body, independent of both parent companies. It will incorporate the strengths of both Fanuc's specialization in controls and GM's specialization in applications. GMF will focus its work on software and sensor developments. The two principals in this venture obviously have a clear understanding of the needs of its end-user and will be able to bid on a majority of GM's business. GMF has already booked orders from GM as well as a few auto suppliers. (See Fanuc company profile for more information on this joint venture.)

Japan also enters these tie-ups to make up for its own deficiencies. Because it is weak in the areas of robot intelligence, electronic controls, and programming software, Japan must enter licensing agreements or obtain through outright acquisition the technology and equipment it needs.

There are three stages in the design and manufacture of AMES systems: (A) product concept and "architectural" design, often patentable; (B) design in implementation or application engineering; and (C) manufacture of systems components. The United States has pre-eminence in (A), and Japanese are strong in (B) and (C). From a competitive viewpoint, although weak in (A), Japanese firms are able to purchase what they need from U.S. sources. Particularly accessible are the smaller, independent West Coast firms such as "Daisy." This firm has been in business for a little over a year, employs about

80 application engineers (many of them spin-offs from larger software design groups, such as INTEL and G.E.'s Colmar Corporation). Daisy sales are now in the $12 million range, about 50 percent overseas to Japan and Europe. Top engineers are attracted to these smaller companies by matched salaries plus attractive stock options. Daisy is now in the process of negotiating a Japanese affiliation which will cost-effectively manufacture in Japan work stations sold by Daisy.

There are now literally hundreds of these spin-off companies that are ready, willing, and able to provide Japanese firms with the missing "A". Another example of this type of access is the teaming up of the Ungermann-Bass (United States) group with Fujitsu (Japan) to jointly develop an "ether network" (communication traffic control) device. UB provided the basic concept and architecture (A), and Fujitsu provided the detailed engineering to design and build the new device. Fujitsu is strong in "gate-array" engineering (coding and decoding systems), which is central to the new device.

AMES Users

Overview of AMES Demand

Total sales of robots in Japan in 1981 amounted to 170 billion yen (approximately $680 million). Most forecasters predict that this figure will climb to 1 to 1.2 trillion yen ($4-4.8 billion) in 1990. Such estimates reflect the growing acceptance of robots in the workplace by both management and labor. Both see themselves as standing to gain by the proliferation of robots in the manufacturing process.

Because of the demand for a greater variety of goods by consumers and the shortening life-span of each model, Japanese companies have turned to AMES to enhance the flexibility of production. AMES gives its users shorter lead times in producing new models and concurrent lower inventory costs. AMES also allows companies to increase their production levels without having to absorb new labor.

The automobile industry has historically been the largest user of robots. But in 1980, the electronic equipment industry rose to first place for the first time. Table 10 gives a breakdown by industry of user demand for robots. Even though the auto industry is no longer the principal user of robots, it still dominates the sphere of sophisticated robots.

The auto industry uses the playback robot for arc and spot welding and spray painting. For the majority of Japanese car models, each unit requires 3,000 to 4,000 spot welds. Programmable robots have replaced multi-spot welders which were designed specifically for each car, making model

TABLE 9

ROBOT SHIPMENTS TO INDUSTRY IN JAPAN

| Industry Sector | (as a percentage of total value) | | | | | Compounded Annual Growth Rate |
	1976	1977	1978	1979	1980	
Auto Manufacturing	30	34	39	38	29	54
Electronic Equipment	21	23	24	18	36	76
Synthetic Resin Processing	13	10	10	11	10	44
Metal Product Manufacturing	6	3	7	8	5	47
Metal Processing Machinery	5	6	4	3	4	46
Steel	6	6	3	4	1	39
Non-Ferrous Metals Manufacturing	2	1	2	2	3	70
Export	2	4	3	2	3	70
Others	15	13	8	14	9	32
Total	100	100	100	100	100	54.0
Total Shipment Value	13.7	21.1	26.6	38.0	76.9	

SOURCE: P. Wolff and H. Matsuzaki, *Japanese Robotics: The Take-off.* Tokyo: Bache-Halsey, Inc., 1982, p. 20; based on statistics presented in *Sangyo Yo Robot Guidebook,* Kikai To Kogu, October 1981, p. 13.

change-overs costly and time consuming. Robots require as little as one hour to be reprogrammed in the event of a new model. Spot welding robots account for over half of the playback robots used in the auto industry.

Arc welding and spray painting are performed in the most unfavorable environments. Workers frequently need to take time out from their work, creating disruptions in production. In addition, they require years of experience to train a skilled worker, and skilled workers are becoming increasingly scarce. The introduction of robots has solved both problems, filling the shortage of skilled labor while they freed workers from hazardous and unpleasant surroundings.

The demand for assembly robots will surely outstrip the demand for playback robots in the near future. Between 1976 and 1980, sequence and playback robots dropped from 84 percent of the market to 61 percent, whereas NC and intelligent robots climbed from a 5 percent to 30 percent share. This latter group is expected to increase their share of the market to 40 percent by 1985 and 50 percent in 1990. The growth of the incidence of use of these intelligent robots means that the average value per unit in use will increase, thereby increasing the total market value of robots marketed.

This move towards intelligent robots is in accord with the government's wish to build up the knowledge-intensive industries. For example, assembly robots, which are used extensively in the semiconductor industry to weld contacts on microchips, accounted for only 15% of robot sales in 1980. Assembly robots are still in the experimental stage of development, but it is predicted that they will claim 17% of the market by 1985 and 22% by 1990. The development of these intelligent robots is also in line with the projected increase in exports. It is expected that foreign markets will demand the intelligent robots more than the less sophisticated, essentially labor-saving robots.

Industrial Management and Labor Relations

Japanese managers (and labor) are generally highly receptive to the introduction of AMES systems. There are several reasons for this. One is that Japanese companies do not have to answer to stockholders. They are not judged by quarterly profits and therefore have a greater propensity for innovation. Management is willing to risk disruptions to production in the short run in order to increase productivity in the future. This long-term view is in sharp contrast to the reluctance of American managers to accept any such disruption. It also explains in large part the success of robots in Japan. The introduction of robots can be very expensive. Increases in depreciation, interest costs, and the miscellaneous costs involved in changing the plant and its equipment to accommodate the robot all drive up the costs. Moreover, interference and slowdowns in production are likely to occur while the robot is being fully integrated into the production process.

One Japanese company experienced a decline in production and a 30% increase in total costs during the first year. However, by the end of the second year total costs were 25% less than if the product had continued to be produced manually.[7] The persistence of the Japanese company paid off handsomely in a situation that would have caused most American managers to abandon robots.

Japanese managers are protected by the same life-time employment system enjoyed by all workers in the large Japanese companies. Instead of concentrating on maximizing present production with existing equipment, Japanese managers are able to seek new, more efficient production methods. Since they are not likely to lose their jobs on the basis of one quarter's or even one year's performance, they are more willing than their American counterparts to assume the risk of technical innovation in the hopes of long-term improvements. (However, within large Japanese companies, managers are often rotated every two to three years, and must show "results" within that time period.)

Labor-Management Relations

The rapid introduction of robots into the workplace has not yet raised fears of displacement in Japanese workers. The life-time employment system of the largest firms guarantees that workers will not lose their jobs due to robots, but will be shifted elsewhere in the firm, often to better positions. Futhermore, since a worker's wage bonus is directly tied to the welfare of the company, any device that improves productivity and, in turn, profitability will be seen as an asset. Robots also free workers from unpleasant or dangerous work and free them to be retrained and transferred to a different part of the company.

It is expected that the introduction of robots into the production process will actually create a new sector of the workforce. These workers will be needed to program and maintain the operation of the robots. However, this may also be a constraint, at least in the short-term. Until there are enough skilled workers to properly run the robots, companies will be hard pressed to rapidly automate their production process.

Robots also allow companies to increase production levels without hiring new workers. Japanese companies are reluctant to absorb new labor during an upswing of the business cycle because of the added burden they create when business drops off. Robots are able to operate on multiple shifts, increasing production output without increasing capital costs.

Labor unions in Japan are organized to include all the workers of a firm, instead of grouping workers with a specific skill, such as electricians or mechanics, which is the norm in other countries. These company-wide labor unions therefore, have a more cooperative, as opposed to adversarial, relationship with management. During times of slow business, workers know that the executives of the company will be shouldering their share of the cutbacks. Also, the wage differential from the lowest worker to the top executive is nowhere near the spread found in the United States. The ratio of the lowest pay to the highest pay for a typical Japanese firm is around 15:1, and an even lower ratio of 8:1 or 10:1 is not uncommon. By comparison, the wage ratio for U.S. firms averages between 50:1 and 70:1, and often exceeds 100:1. For these reasons, labor tends to identify more with the company's viewpoint. This substantially decreases labor's opposition to innovation.

There are several factors that explain labor's acceptance of automation.

First, under the wage structure in Japan, most workers are given a bonus twice a year, which, depending on the success of the firm, may be equivalent to several months pay each year. Since robots enhance productivity and therefore its profits, workers share in that success.

Second, robots have taken over the most dangerous and repetitious jobs, thereby "humanizing" the workplace. Workers who are shifted into more interesting and challenging positions also welcome the opportunity to work in safer and cleaner environments. Given the fact that workers are trained in several different skills, there is no problem transferring these labor generalists to other parts of the company.

Third, workers in Japan have long been committed to quality control. The first Quality Control (QC) circles were formed in the mid-1960's. These QC circles, usually consisting of 10 or so workers, were responsible for improving the level of workmanship in their area. Robots are seen as yet another means of improving quality control.

Finally, the higher level of education and training of the Japanese worker removes the fear of new technology. Japan has a long history of factory innovation. Robots, which are the newest example of this innovativeness, are not resisted by labor.

Although labor has not challenged the introduction of robots, there are several problems which will have to be faced in the near future. The life-time employment is found only in the largest of Japanese companies, employing roughly one-third of the labor force. These industrial giants were the first to utilize robots, and the workers of these companies had no fear of displacement. But as robots begin to spread to smaller firms, which do not guarantee life-time employment, the remaining two-thirds of Japanese workers may become victims of automation.

There is at present a shortage of skilled labor and demographics suggest the labor force will grow only 0.7% per year for the next several years. It is hoped that robots will fill this skilled labor shortage. But if the labor force does begin to expand again, entrance to low-skilled jobs will be blocked by the presence of robots. Consider also that most skill training is done by the company, which may be reluctant to invest in the skill training of inexperienced, unproven young people.

Management-Labor Receptivity to AM Systems

As the production of robots has progressed and improved, the costs and risks of introducing them into the workplace have greatly diminished. Whereas in the early 1970's, the cost of a robot as a ratio of the average wage was 10.5, following the oil shock of 1973 and the resulting increase in wages, the ratio at present has been reduced to 3.1. Not only has the relative cost of a robot declined, but the absolute cost has decreased as well. The lower price of robots today is due to the lower costs of micro-chips and other hardware and because robot producers have moved farther down the learning curve, thereby lowering prices.

Favorable government policies also enhance the introduction of robots. Firms utilizing robots are entitled to depreciate 12.5% of the purchase price during the first year, in addition to regular depreciation. The establishment of JAROL has also allowed firms to lease robots without a large initial investment. A robot valued at 6 million yen leases for five percent down-payment and about 100,000 yen per month over a six year period. (The equivalent wage of a machine operator is about 140,000 per month.) Interest rates run about 7.5 to 8.0 percent per annum.

Because the cost of a robot has decreased so markedly during the past several years, the time needed for the payout of a robot has shrunk accordingly. If a robot replaces one worker for one shift per day, the payout is roughly four years. If the robot is used for two shifts, it will pay for itself in two years and if used round the clock, in just over one year.

There is a general reluctance to lay off factory workers during the downward trend in business cycles. The introduction of AMES is viewed by a broad cross-section of Japanese firms as an effective means for coping with fluctuation in the business cycle.

It is the current shortage of skilled labor that perhaps has had the most to do with the proliferation of robots. There appears no end in sight for this shortage. At present, the demand for skilled labor outstrips the supply by over 8 percent, a figure comparable to the unemployment rate in other countries. Robots quickly fill this gap by requiring only a few minutes to program, rather than the several months or years to train a machine operator.

In addition to the declining costs and risks of utilizing a robot, there are considerable gains to be had. Robots improve productivity by being tireless workers and working under hazardous conditions. Robots are able to do the most boring and repetitious work without tiring or becoming sloppy in their work, which guarantees quality control. They do not require heating or air-conditioning, ventilation or lighting, resulting in savings of production costs. Robots can also cut costs by the conservation of materials. Robots used for spray painting use up to 30% less paint than their human counterparts. The durability of robots has improved to the point where there is very limited down time.

The demand for software and programming in robotics is outstripping supply of applications engineers. The larger companies, such as Hitachi and Toshiba, are better positioned to recruit and train design engineers, but second-tier companies, such as Sharp and Casio, have much greater difficulties in obtaining the necessary talents and skills. One expert in the field estimated that 80 percent of the companies in the AMES field are "starved" for critical personnel. A major deficiency lies in the educational system, which is very weak in the basic sciences underlying the new design-engineering requirements. There is an over-supply of technical people in the steel industry, but these people cannot be "retooled."

At international technical meetings, Americans often present frontier ideas and concepts, which the Japanese are quick to take note of and follow up on. (Japanese technical papers, in contrast, report on current practice.) Japanese firms will often disclose operational plans to elicit outside criticism—as part of their efforts to fill in missing details or to detect significant flaws in their thinking. Cases have been known where the Japanese firm invites an American firm (with a new product concept and seeking capital resources) to make a presentation in which enough detailed information is elicited so that with a modest additional intelligence effort they can strike out on their own.

But it is the flexibility that robots provide that are their most attractive feature. The whims of the marketplace have shortened the life cycles of many products and have forced producers to provide new models at an ever faster rate. Robots can be quickly reprogrammed to perform another task. Since they are not designed for one specific task, robots can be used during the life cycles of a variety of products. In a highly competitive market such as Japan, it is essential that producers be able to meet the constantly changing demands of the consumer public. The flexibility of robots allows companies to respond immediately to changes in the marketplace.

Because robots are but one part of a greater, centrally controlled and automated manufacturing system, they can be used for shorter production runs, as well as provide efficient transporting of supplies and materials within the factory. Also, less lead time is involved in getting a model change into production, and inventory costs are cut because the flexibility of robots allows the production process to be rapidly altered to fill new requirements.

Even though the costs and risks of robots are greatly outweighed by their benefits, there are other considerations which suggest that the robot industry may not experience unlimited growth. A robot can conceivably perform 24 hours per day and thereby hasten its payout, but there is the question of whether or not there is the *need* for round the clock production. Especially with the current slowdown in domestic orders and the problem of declining exports because of the world-wide recession and resentment of Japanese trade practices, there may be no market for increased production. Moreover, because of the problems of acquiring materials, there may not even be the *ability* to produce 24 hours per day.

The intrinsic nature of the robot industry itself may slow future growth. One of the best features of robots is their flexibility and their consequent ability to be quickly reprogrammed. Thus, companies may find it unnecessary to scrap their present robots for newer ones since the original robots will not soon become obsolete.

Technical Absorptive Capabilities

There are several factors which enhance the capabilities of Japanese firms to absorb technical innovation. One is the lifetime employment system. As described in previous sections, the lifetime employment systems typical of larger firms greatly reduce the fear of managers and workers that they will be replaced. This means that managers are able to concentrate on improving the methods of production instead of merely increasing production using existing technologies. They are more willing to fine-tune the production process and experiment with new technology, secure in the knowledge that short-term disruptions and isolated mistakes will not cost them their job. Workers, for their part, do not fear being displaced by technical innovation, such as the introduction of robots. For the most part, they welcome the opportunity of having robots take over the dreary and dangerous chores so long as they can be transferred within the company. Since any improvement in the fortunes of the company show up in their wage bonuses, workers do not resist a new technology which improves productivity.

Japanese firms are not willing to settle for prevailing "mature" technologies if they see a possibility for improvement. The ability of Japanese firms to improve on technologies developed elsewhere is the key factor which has fueled the rapid development of their economy during the last two decades. This is especially salient in the development of the robot industry in Japan. Although robots were first produced in the U.S., Japan today accounts for 70 percent of all robots produced and in operation throughout the world.

It is important to restate here that most Japanese producers of robots began producing robots for internal use. The development of the robot industry was done to improve productivity and only much later was the issue of commercialization considered.

The relatively large numbers of engineers and other technical personnel employed by Japanese firms also increase their ability to absorb technical innovation. On a per capita basis, Japan graduates twice the number of engineers as does the United States. Their presence in both the design and production sectors of a firm means that the technical know-how to evaluate and institute new technology is readily available.

Finally, the highly competitive nature of the Japanese market means that industrial firms are all but forced to seek new innovations. In order to keep their competitive edge, Japanese firms must continually find ways to keep costs down by tinkering with the production process. The introduction of robots results in reduced labor costs and improved productivity and quality control, and their flexibility allows user firms to keep abreast of changes in market demand.

FOOTNOTES

[1]Drawn in part from Jack Baranson and Harold B. Malmgren, *Technology and Trade Policy Issues, op. cit.*

[2]See Jack Baranson, *Sources of Japan's International Competitiveness in the Consumer Electronics Industry*, prepared for the Office of Technology Assessment, U.S. Congress, June 1980.

[3]Japanese industrial planning mechanisms set targeted dates for moving to next generation technology in the VLSI area. This involved progressively moving from 4k to one million k RAM (random access memory). In the floppy disc field, used in office and home computers, it involved reducing the size of these discs from 8 inches to 5¼, and more recently 3½ inches. The line-widths in VLSI (very large scale integration) devices are critical, and Japanese industry has set goals to move to 6 micron to 1 micron tolerances.

[4]The major technical advances planned for the seven-year effort are: a) improved sensory capabilities, visual pattern recognition, force/torque sensors; b) control algorithms incorporating the adaptive or "intelligent" behavior, eventually intended to allow the robot to operate more or less automatically, making decisions based on the sensory data it receives; c) mechanical design, including developments in manipulators, and capability for locomotion (the latter very much a controls problem as well).

[5]See Part III for profiles of Japanese firms.

[6]Detailed information specific to Japan's leading robot manufacturers can be found in Part III. See Table 8 for overview.

[7]See Paul Aron, *Robots Revisited: One Year Later*, Paul Aron Report No. 25 (New York, New York: Daiwa Securities America, Inc., 1981), p. 11.

CHAPTER VI
EUROPE

National Environment

Government Industrial Policies

West European countries are more heavily dependent upon exports than the United States and also import more in relative terms. Because of the greater weight of exports in national income earnings and the need to compete in international markets, technological innovation (including automation) and increased productivity are major goals of West European governments. Most West European governments have addressed this need by formulating national policies which are specifically directed towards robots and computerized manufacturing processes. These national policies, however, vary greatly, ranging from France's designation of the robotics industry as a key sector for future public investment to Sweden's lack of specific measures aimed at AMES industries.

France

In France the Committee for Development of Strategic Industries has identified industrial robots and automated production systems as a key sector of the economy. Basically, the government has two goals: a) to encourage the manufacturing industries to invest massively in automatic equipment (so as to increase productivity and thus be able to improve their share of the market both in France and abroad) and b) to facilitate the development of a strong French industry in the AMES sector. A key factor is that French industry has declined to eighth place worldwide in the machine tool industry.

The government's first major plan was developed for the machine tool industry. The Machine Tool Program provided for doubling machine tool production, reducing imports by half and increasing exports over the next three years. Four billion francs will be spent to achieve this goal. The government will provide 2.3 billion francs as direct aid while the remainder must be raised through loans. The program is supposed to end the industry's seven-year crisis and make it competitive again. The government's objective is to specifically place 16,000 new computer controlled machines in service between now and 1985.

Robotics has been designated as a key industry for public expenditures on R&D. The advanced Automization and Robotics Group, consisting of ten research institutes, was formed in October 1980. The project will run through 1984, and the group will emphasize general robotics, engineering and technology, advanced remote operations and flexible production systems. Total expenditures over a three-year period will run approximately $400 million (2.4 billion francs). Supply factors as well as demand factors will be acted upon and the Mitterand government plans to increase its procurement of robots (now 50 million francs to 1.2 billion francs over a period of three years). Developmental contracts have already been issued by the Committee for the Development of Strategic Industries (CODIS) to the following French manufacturers: CEM, Prodel, Trefilerics and Atelier de Commercy, Afma-Robots, CFC (Compagnie Francaise des Convoyeurs), and Automatique Industrielle. The government has set a goal of 5,000 robots and the establishment of 410 research positions in three years, and the creation of two thousand qualified jobs over the next ten years.

Among the measures advocated for developing the robotics industry is one to restructure the industry around some of the largest firms. France currently imports over 60 percent of the robots it uses. The density of robots in use in France also is relatively low: only 0.3 robots per 10,000 industrial workers as compared to 11.2 in Sweden, 4.4 in Japan, and 1.6 in the United States.

Restructuring measures reflect a policy change from one of stimulating demand to one that brings together the user industries, the producers, and the government in a comprehensive program. Included here are efforts to identify potential areas where robotics could be introduced coupled with a consideration of measures to facilitate acceptance of AMES systems (including the retention of displaced workers).

Production currently has many deficiencies, particularly in the basic components sector. The government's goal is to complete a "robotics industrial chain." The supply end of the chain will be restructured to facilitate the regrouping of the marketing activities of the small producers. The aim is to restructure the robotics industry around a few of the largest enterprises, such as Renault's subsidiary ACMA, whose activities are currently oriented chiefly towards the automotive industry.

Sweden

Sweden has little in the way of a concerted policy towards the robotics industry. Of the countries studied, it has the most *laissez faire* attitude. The Swedish government is not as concerned with the development of robotics as it is with its application and integration into production systems.

Government financing of research and development has had a very limited influence. In the period 1972-1979 the Swedish Board for Technical Development (STU) allocated only about $800,000 for the development of the robotics industry. But during the 1980's, STU's support of research and development in engineering industries will increase significantly to about $40 million. STU, under the Ministry of Industry, is funding advanced research and development in universities, research laboratories, and industry. Funding for robotics and CAD/CAM has greatly increased in recent years and about $8 million was allocated for CAD/CAM during 1980-1985. The Swedish Government is also proposing the establishment of three CAD/CAM centers as a joint venture between universities and government engineering development centers. In addition, there is an agreement between STU and the Swedish Association of Mechanical and Electrical Industries to sponsor a five-year research program through 1985.

The Computer and Electronic Commission thus recommended that the government promote a wide diffusion of robotics and CAD/CAM through: a) an information campaign among small- and medium-sized firms; b) complementary software development loans with conditioned repayment; c) encouragement of the vocational training program at the Swedish Institute for Corporate Development; and d) various related training programs.

West Germany

In West Germany, government and industry support for manufacturing technology development (of which robotics is a significant part) is estimated at $100 million annually. The Federal Ministry for Research and Technology (BMFT) sponsors a number of programs with involvement in production and manufacturing areas and robotics. The BMFT also funds the Association for Fundamental Technology (GFK) which is involved in development programs in computer-aided design and process control by computer; and the German Institute for Aerospace Research and Experimentation (DFVIR) which coordinates and monitors "Humanization of Working Conditions" and "Advanced Manufacturing Technology" programs, as well as developing sensors and feedback systems for robots.

The German Research Society (DFG) supports research on machine tools and controls, robotics, production systems, production engineering and manufacturing technology. About $30 million a year is provided for these programs, half from the Federal Government and half from the States. An organization of institutes for applied research, the Fraunhofer Foundation, currently has two members, the Institute for Production and Automation (IPA) and the Institute for Data Processing in Technology and Biology (IITB), who are involved in industrial robot developments.

Italy

The Italian government gives very little aid to robotics research. The

Institute of Electro Technics and Electronics, Polytechnic School, Milan, is involved in a number of studies including stepping motor performance in a robot, multi-microcomputer structures, programming language, object recognition and programming assembly problems. Several other universities are also participating in specialized robotics research projects.

Unfortunately, very little of the research and development work in industry is publicized. One exception to this is Olivetti which manufactures relatively small, high-capability robots for assembly.

AMES Producers

Overview of Industrial Structure

In Western Europe production capacities vary greatly among major AMES producers. Sweden, which began using robotics in 1970, and West Germany have more extensive and developed production capacities than those of the other West European producers. Sweden and West Germany have concentrated on exporting their production while other West European countries, France, England, and Italy, have imported industrial robots from the United States and Japan and have yet to develop extensive exporting.

West European AMES producers have been influenced by external market penetration more than by domestic or intra-European competition. This is due in large part to the AMES industry structure in Western Europe. West European producers of AMES have emphasized performance over quality, allowing Japanese firms to undercut their prices. West European firms, notably ASEA, are now attempting to gain a larger market share, domestically and globally, through the scale advantages of increased production.

While the number of West European AMES producers is increasing rapidly, the industry structure is dominated by a few major firms. ASEA (Sweden), Volkswagen (West Germany), Trallfa (Norway), Renault (France), and Olivetti (Italy) are the major producers. These producers are characterized by their large size and established position in closely related market areas such as the automotive industry (Renault, Olivetti).

Sweden

In Sweden, the AMES industry is characterized by three types of producers:

1. Manufacturers of a line of *multi-purpose* industrial robots. The main firms within this grouping are ASEA, Atlas-Copco and Kaufeldt. Their combined production for 1982 is estimated at 1,100 units (of which ASEA will produce over 1,000).

2. Manufacturers of *special-purpose* robots that are custom-designed to service specific machine tools. Volvo's "Dappen" for loading/unloading press machines is an example of one which has been sold worldwide to automobile manufacturers.

3. Manufacturers of *programmable material handling* equipment (e.g., automatic transfer systems, computer controlled cranes, and warehousing systems). Sweden has a strong international position in this sector of the AMES industry. The leading companies in this area are Volvo, BT Lifters, Tellers, Digitron, ASEA and Moving. The equipment produced by these · manufacturers is integrated into flexible manufacturing systems produced in Sweden.

The domestic market remains important for Swedish robot producers since it can act as a testing ground for new robot systems, and user demand in Sweden is more sophisticated than in many export markets.

Italy

Currently thirteen companies are producing industrial robots in Italy, an increase of nine since 1972. Italian robot producers can be classified into three major categories:

1. *Subsidiaries of large user industries* (e.g., Comari for Fiat, supplying mainly flexible manufacturing systems, and Osai for Olivetti.)

2. *Medium-sized companies* which have traditionally manufactured measuring instruments or fixed welding and painting stations and have diversified into robotics: e.g., Basfer, Gaiotto, DEA, Elfin, Bisiach e Carru and Morin:

3. *Small,* recently created firms specializing in a particular segment of the AMES field; e.g., Norda, Aisa, Camel, Jobs and Ses.

Italian robot production has increased dramatically during the last few years with an average growth rate of about 60 percent annually. Much of this production was exported; in 1979 420 robots were produced and 160 (or 38 percent) were exported. According to industry association estimates there are 750 installed robots in Italy, with the largest percentage in spot welding (16 percent), followed by spray-painting robots (12 percent). A limited number of assembly robots are used in limited electronic and electromechanical component factories.

Fiat has led the field in the automation of production processes. It developed the Robogate system which allows the assembly of several different models on a single production line. Other Italian AMES producers have concentrated on developing new applications and on improved robot quality rather than on increasing robot production. Robots are manufactured in small lots as producers attempt to respond to changes in user demand. The R&D budgets of most Italian AMES producers are devoted primarily to the improvement of product design applications, such as improved sensors for arc welding.

Italian robot producers have not only dominated their domestic market but also have a relatively strong export position. The Olivetti Sygma robot for assembly, and the Basfer painting robots, have had sizeable export success. DEA has recently granted General Electric a license for its Pragma assembly robot.

Italian producers of robots, however, have one major problem: they depend strongly on foreign suppliers for components which makes them somewhat more vulnerable to market uncertainties.

France

France was slow in coming to realize that it was trailing its competitors in the production of robots. But of late it has set ambitious goals for itself so as to not be left behind. Government and industry have combined to set a target figure of 5,000 "true" robots to be met in 1990. This will be dificult to achieve because of the small number of robot producers in France and the already substantial penetration by foreign producers. Sixty percent of the French market is claimed by imports. France hopes to cut this to forty percent by 1990.

France's robot producers are at a distinct competitive disadvantage. They are not only off to a late start, but they are making their move when the domestic and international economies are in a prolonged slump. Many of France's small producers of numerically controlled machine tools have folded in the face of strong international competition, despite substantial government subsidies. Other larger firms are finding it difficult to devote adequate funds to research and development of new models. Even France's largest robot producer, the Renault subsidiary ACMA, is only one-tenth the size of Unimation, the world's largest manufacturer. France is committed to improving its competitiveness, but it will be some time yet before its relative market position can be determined.

Corporate Strategies and Outreach

Most European robot producers made the same mistake in entering the market that the United States did: they started too big. Both European and American firms began by producing large, complex, and expensive robots. As a consequence, they were easily undercut by Japanese firms that offered smaller and cheaper robots to foreign users who were interested in automation, but wary of investing large amounts of capital in untried equipment. European producers have since realized their mistake and are beginning to scale back their production.

Most European producers in general are not as large as their Japanese and American counterparts. They cannot devote the same amount of capital to R&D, and they are also at a disadvantage because they are being forced to catch up to their foreign competitors. Also, European firms have long emphasized performance in their produced equipment and therefore find it difficult to compete with cost-efficient, high-quality Japanese robots.

One exception to this general trend is the Swedish company ASEA. Only 10 percent of its sales are to the domestic market. ASEA's strong product efficiency has enhanced its determination to maintain a worldwide position. Its high production rate has allowed it to produce more cost effectively and consequently to increase its growth rate and market share. ASEA claims 15-18 percent of the global market (excluding Japan) and around 50 percent of the Swedish market. It is able to spend funds necessary for its R&D—about $200 million last year. It conducts its research in close cooperation with its customers in order to meet their needs properly. It provides a wide array of robot models for maximum visibility and competitiveness, and backs this up with a marketing and servicing network to its customers.

Swedish firms in general place a high priority on engineering in their corporate strategy. Engineering accounts for nearly 50 percent in terms of production and employment in manufacturing industries. The distinguishing feature of Swedish manufacturers, as opposed to other industries, is the low rate of fixed capital goods in contrast to the high rate of investment in R&D, which promotes the knowledge-intensive industries.[1]

Product Spread and Market Share

French producers of robots are few in number and small in size. Because they do not have the resources to develop fully their robot production, large gaps exist in the supply of domestically-produced robots. The notable exception is Renault, which entered the market in the mid-1970's. It devoted $15 million and five years of effort to enable itself to enter the field, concentrating its production on the top and middle of the line. Renault is the only French producer offering a full line of robots and the only one with intelligent robots. It has the ability to export to the United States and Asia, and by the mid-1980's should be a major world producer. The only significant competitor to Renault is AOIP, which is a former workers' cooperative. It has only recently joined the field, producing 15 units in 1980 and 40 units in 1981.

Among Swedish producers of general purpose robots, ASEA dominates. Its production target for 1982 is 1,000 units, making it one of the three largest robot manufacturers in the world. ASEA is currently trying to increase its market share worldwide by reducing the price of some of its robots by 25 percent, taking advantage of scale economies from large production levels. ASEA exports 90 percent of its production and views the international market as essential to its success (as is true for other West European AMES producers). As one means to more effectively penetrate foreign markets, ASEA has set up production facilities in the United States and Spain.

The largest share of the Swedish domestic market for AMES at the end of 1981 was held by ASEA-Electrolux (26.5 percent), followed by Kaufeldt (22 percent), Trallfa (10 percent), and two U.S. firms, Unimation and Cincinnati Milacron (9 percent and 8 percent, respectively). Other minor Swedish producers are Ekstroms Industri AB, Torsteknik AB, Atlas Copco, and Hiab-Foco.

Joint Ventures, Licensing Agreements, and Acquisitions

Japanese penetration of West European markets was a major factor in the development of licensing agreements between European and Japanese companies. (See Table 10.) European dependence upon foreign suppliers of AMES components is one of the reasons for joint ventures with Japanese firms. These new partnerships also include technical interchange, cross-licensing, joint research efforts, and joint marketing arrangements.

Some of the joint ventures are among European companies. Trallfa (Norway) for example has teamed up with several Italian AMES manufacturers that design and build flexible manufacturing systems. This is one way to expand sales of their painting robots.

AMES Users

Overview of AMES Demand

Demand for AMES in Western Europe is expected to increase significantly within the next decade. West European robot sales totalled 1,700 units in 1980 ($95 million) and are expected to total 8,178 units by 1986 ($558.6 million). The total number of robots in use in Western Europe as of 1981 was over 5,000 with West Germany and Sweden the leaders. Alone, West Germany accounts for 2,300 units followed by Sweden's 1,700.[2]

The economic recession, cost of capital and the subsequent reduction or postponement in investment plans have all served to slow the demand for AMES products. But as the economy begins to improve, demand is expected to pick up. Unimation (United States), now supplies 23 percent of the

TABLE 10

JAPANESE-EUROPEAN COMMERCIAL LINKAGES*

Japan	Western Europe
Hitachi	Martin et Cie (France)
	Zeppelin (West Germany)
	Oerlikon (Switzerland)
	Lansing Robots (Great Britain)
Yaskawa	Tors Technie (Sweden)
	Messer Briesheim (West Germany)
	GKN Lincoln Electric (Great Britain)
Shinmeiwa	Grundy Robotics Systems (Great Britain)
	Commercy Soudure (France, Benelux)
	AGA Welding Engineering (Scandinavia)
	Robel Roboter Technik GmbH (West Germany)
Fijitsu Fanuc	Siemens (West Germany)
	Manurhin (France)
	600 Group (Great Britain)
Nachi Fujikoshi	Kuka (West Germany)
	Ayo (France)
Osaka Transformer	ABS (France)
Kobe Steel	Trallfa (Norway)
Dainichi Kiko	Fenwick (France)
Sankyo Seiki	CGMS (France)
Kawasaki	ASEA (Sweden)**

SOURCE: J. Le Quemont, *The Social and Economic Stakes in International Competition in the Robotics Industry,* Agence de l'Informatique, Université de Paris-Sud, 1981, p. 27.

*Includes technical and marketing agreements as of 1981.

**Under discussion.

European market, followed by Norway's Trallfa and three Swedish firms. Europe clearly lacks a strong indigenous robot manufacturing capacity in comparison to Japan and the United States. Western Europe imports 45 percent of its robots, and the United States 50 percent and Japan virtually none.

Swedish industry began to invest in robotics in the late 1960's. By 1970, the number of robots installed was about fifty, but this figure grew at an average rate of 37 percent per year during the 1970's. Sweden's projected growth rate for robots in use during the 1980's is 25 percent. This slower growth rate reflects the fact that growth will be calculated over a larger installed base of robots; that the "easiest" installations will have been completed; and that new alternatives to the general purpose robot will become more attractive.[3]

France has been slow in moving into the era of automation. Nearly two-thirds of its enterprise have no automated equipment and its machine tool inventory has an average age of fifteen years. The French government has set a goal of encouraging industries to invest heavily in AMES equipment (including, but not exclusively, robots). Automation should strongly bolster productivity and improve the market share of French manufacturers both at home and abroad.

AMES demand in Western Europe is accounted for largely by industries which are already highly automated. Since 1977, for example, 70 percent of firms (largely automotive) already using AMES in Sweden have increased their demand for robots. It is the automotive sector which continues to provide the most notable increase in the use of industrial robots (except for Sweden, where other metal working industries dominate).

In terms of application of robots in Western Europe, spot-welding and paint-spraying robots are predominant. General handling, machine loading, and arc welding are the next most popular applications. Assembly and sensor equipped robots are expected to constitute the largest percentage increase in the next decade with assembly robots accounting for 25 percent of the total robot population in Western Europe by 1986. In West Germany alone, the demand for assembly robots will grow from four percent of the present total to ten percent by the mid-1980's, to thirty percent by the early 1990's. Arc-welding applications are also likely to increase given the recent improvements in robot technology within this field.

West European countries are also increasing their research and development of CAD/CAM systems. Work in developing CAD systems is now reaching the commercialization stage. Demand for CAD/CAM technology is centered in large market-leading companies which have the capital resources to utilize the systems. CAD/CAM application is largely in engineering/design analysis, automated drafting, and NC-tape preparation. Industrial computer users, however, still spend more on business information and financial systems than on engineering or CAD/CAM systems.

Small firms account for the bulk of demand in the United Kingdom, however, and the majority of industrial robots there have been installed singly or in pairs. French domestic demand, although still small, was strongly affected in the mid 1970's by the entry of Renault which invested 15 million dollars and five years of effort in order to enter the AMES field.

Role of Competition

Considering Western Europe's economic reliance on exporting, international competition is a significant and growing factor in the adoption and diffusion of AMES. Important sectors of manufacturing in many countries of Western Europe have been surpassed by foreign competitors who are robotizing at a much faster rate. Many firms (and in some cases, entire industrial sectors) will spend a decade simply catching up to where their foreign competitors are now. Firms that do not or cannot automate fast enough in response to international competition will fail, thus reducing effective demand for AMES and perhaps further weakening the competitive position of entire countries.

International competition is especially significant in the automotive industry which is now being automated in many West European countries. Productivity differences may be eliminated in this sector but increase in other sectors of West European economies where diffusion of AMES is slower. Product diversity and production flexibility will increase in importance while other factors such as product standardization and economies of scale will decrease.

Commercial competition (both intra- and extra-European), especially in the lower-cost end of the robotics market, will continue to reduce the cost of AMES, thereby increasing the rate of diffusion of robotics in Western Europe. This trend of price competition is less significant for firms that are producing AMES primarily for internal use (Volkswagen). Yet many European AMES producers recognize the need for outside (and export) sales. Price competition will give rise to other forms of competition in servicing, adaptive design and marketing—all of which increase the rate of commercialization.

Robot equipment costs have already been reduced by commercial competition. ASEA recently lowered the base price of some of its robots from $85,000 to $65,000 in an effort to gain a larger market share. ASEA also claims that it will invest as much for marketing and application techniques as it will for technical product development.

The worldwide recession has reduced capital investment in general, and factory automation in particular. It also has dampened worker and management resistance to automation; both recognize the need to automate in order to remain competitive and many firms are doing so rapidly in spite of the recession. This process causes friction where management attempts to introduce technological change without prior consultation with labor. Such friction inevitably reduces effective demand for AMES as it slows the pace of factory automation.

Where the bargaining power of organized labor is strong (as in Northern Europe), resistance to automation may be a strong factor inhibiting commercialization of AMES. With West European unemployment at unusually high levels (nearly 8 percent in West Germany in 1982), resistance is likely to become more strident. Well-founded mechanisms to involve labor in the process of factory automation (as in Britain, Sweden, West Germany), combined with programs that educate workers on the need for and benefits of automation, will mitigate labor resistance.

The ability of governments and firms to retrain workers displaced by automation is important. Well established procedures for such retraining (especially in Northern Europe) may suffer in the current recessionary economic climate, as governments and enterprises tighten spending. The bargaining power of organized labor is reduced in these circumstances as well. On balance, these currents combine to make labor's opposition a factor that will continue to inhibit commercialization of AMES in Europe in the near term.

Industrial Management and Labor Relations

Industrial management has been most receptive to the introduction of AMES equipment in Sweden, followed by West Germany. France, on the other hand, has been the most conservative in its willingness to automate. The main barrier to the wide diffusion of robots has been the question of how to assess the prerequisites for profitable investment in AMES equipment. A related, but less important barrier, has been the difficulty of raising capital for investment.[4]

Sweden's success regarding the implementation of AMES is based on two factors: first, the realization that technology is important but not as important as good manufacturing equipment; and secondly, good industrial relations, which have been the key to progress. The essence of the Swedish approach (and the Japanese, for that matter) is the belief that driving the most sophisticated machinery hard is not as effective as using adequate machinery 24 hours a day.

High wages in Sweden have been and will continue to be a major factor in leading to the initial decision to automate the production process. ASEA, for instance, has found it to be most profitable to run regular unmanned shifts. This has been particularly true of late, due to the decreased workload which has eliminated the need to hire workers for extra shifts. ASEA's workers have been shifted to other areas in its plants where the work cannot be automated so easily. Workers have also been retrained to operate the robot machining groups. Workers have not been displaced by robots; instead ASEA has achieved reductions in its workforce through attrition. Although displacement has not yet been a problem, neither have there been new job openings. This closing off of entry level positions may become more acute in the future.

France has been the most hesitant to commit itself to automation, but it has now been forced to take a more progressive stance. In 1976 the Sahl Leduc company acquired its first NC lathe. By 1981, it had 12 NC machines and also a robot machining cell for turning—the first of its kind installed in France. Its goal has been to improve quality control, reduce its costs, improve productivity, and raise the level of responsibility of its workers. Automation in French companies will likely become more prevalent in the future in order to comply painlessly with regulated work hour reductions and still maintain (or increase) production output.

There is the growing realization among European countries that if small companies are to meet future demands for increased efficiency and speed in the production process, automation will be essential. The development trend towards increased automation will require engineering and investment efforts far beyond the resources of small companies. The ideas and need for automation are there, but the ability to bring those ideas to fruition is not. There will be a need for efficient use of qualified technical personnel, perhaps by cooperative arrangements between firms. Automation will also require heavy financial support by the government. Not everybody is cheered by the prospect of government assistance to small companies; already in Sweden there is the feeling among some economic authorities that if small companies are given financial assistance to automate, it will result in a distortion of the market.

Cost-Benefit Tradeoffs

In determining the cost-benefit tradeoff of automation, there are usually three factors which must be considered. First is the level of general technology and engineering know-how in the company. Without this, firms (especially small firms) cannot begin the learning process, which is the awareness of new technology, the recognition of the benefits automation can bring and, the ability to undertake feasibility studies and investment calculations. Secondly, knowledge in assessing the prerequisites for profitable investment in automation must be taken into account—not only the initial costs, but also the payback, present and future production levels, and the capacity for technical absorption of AMES equipment. And thirdly, the lack of skilled labor, which is both a barrier and an incentive to invest in robots and other AMES systems, must be considered.

Although progress towards factory automation in Western Europe has been hindered by management risk aversion and fear of change, substantial progress has been made in terms of management attitudes. Reduction of labor costs, increases in productivity, improved product quality, elimination of dangerous jobs, and increased product flexibility are the main factors, ranked in decreasing order of importance, explaining West European management's willingness to invest in AMES. One French manufacturer automated to avoid sub-contracting for reasons of confidentiality.

A shortage of skilled workers was for many years a factor that encouraged European management to automate factories. This trend continues even while much of Europe moves into a labor surplus situation. Wage rates have remained high and the savings in direct labor costs provided by robots outweigh their high investment cost. In management's view, the risk of investing in AMES has decreased as equipment becomes more reliable and flexible as well.

Although the number of flexible manufacturing systems in Western Europe is only twenty-five, the increase in productivity has been significant. The Citroen flexible automated workshop in France yields a productivity 5.8 times that of a traditional workshop of the same capacity. Performance has increased as well in other fields: reduced fabrication times, lower production finance requirements, and earlier recovery of investment.

In the United Kingdom, the industries adopting AMES most rapidly are those with low productivity growth such as mechanical engineering, aerospace, metal fabricating, and instrument engineering. Approximately 40 percent of engineering products in England are produced in batches of fifty or less. Robotics have aided small-batch production techniques by making small production runs economical and by increasing the rate of throughput substantially.[5]

Italian automakers, notably Fiat, are steadily increasing their use of robots in the production process. Fiat has grouped robots in machining centers and is using induction controlled driverless transporters which maintain a flexible material flow. This automated equipment is being utilized in engine building, body production, and in the assembly of mechanical parts. Production can now be matched with great flexibility to changes in demand. Fiat plans to further increase its use of robots (by a third or more) in order to further expand automated spot welding.

Volkswagen, the major automaker of West Germany, plans to follow the example of Fiat by increasing robotized production in its factories. Currently, there are about 570 robots at work in the various Volkswagen factories, and

Volkswagen has plans to increase this number to about 800, mainly in welding and handling operations.

The Swedish automaker, Volvo, best reflects the efforts that European manufacturers have made to incorporate automation within their manufacturing operations. Sweden is the European leader in the development of industrial robots and has installed more of them per capita than any other country. (See Table 7.) Computer controlled machine tools and robots are an integral part of the production process in Volvo factories. ASEA and Electrolux robots trim, grind, weld, and assemble parts, moving the parts automatically from machine to machine. In the most modern part of the factory, five computer controlled machine tools work simultaneously on gear parts, while a single Unimation Puma robot ensures that the parts are passed from one machine to another in the correct order. One man can look after three to five machines, four continue running while he maintains and adjusts the fifth. The advantages for Volvo include reduction of inventory by twenty percent, and savings in interest charges on working capital.

Labor Management Relations

Western European labor's reaction to automation has not included significant resistance to technological change in the manufacturing sector, although there has been controversy over terms and manning levels. Both labor and management concur on the desirability of increasing the size of the pie, but disagree as to how it should be divided. Just as government policies on automation vary among West European countries, labor has also developed different approaches to automation.

Labor relations in some European countries are centralized and include codetermination. In countries such as West Germany and Sweden, progress has been made in the establishment of codes of practice for the use of automation. Sweden has a legal framework whereby employers contribute to the costs of "worker consultants" who advise the unions on complicated technological changes. Sweden's system of codetermination has adapted to technological change readily. A recent policy paper from the metalworkers' union states: "Investments in new technology must increase. . . .there is a risk that Swedish companies cannot meet the competition from other countries if they cannot keep up with the technological development."[6] Swedish unions recognize the benefits of factory automation through the elimination of dangerous jobs and the provision of better tools for skilled workers.

Sweden's history of labor/management codetermination goes back to 1938. This policy agreement was extended to more advanced kinds of manufacturing equipment in 1982. Codetermination has both legal force and popular support, and thus eases the way for increased effective demand for AMES. However,

Swedish labor unions have shown a tendency to rely on labor legislation rather than negotiations with management to deal with problems that arise.[7] This tendency is compounded by management's habit of strictly abiding by the letter, not the spirit, of the law.

West Germany also has a system of codetermination that includes work councils which exist by law in all but the smallest firms. These work councils discuss technological change in the workplace. As unemployment grows, there is increasing dissatisfaction with this system, which keeps discussion of new technology in the consultative mechanisms of the work councils and outside the system of actual negotiations. IG Metall, the large metalworkers' union, has a partly government funded experimental team of advisers on automation designed to assist local unions with problems arising from the introduction of new technology. This project is part of the federal government's "humanization of work" programs.

In Italy and France there has been little done other than talk about the effect of factory automation on workers. In Italy the 1979 round of contract negotiations produced wide-ranging acceptance of clauses requiring management to disclose plans for the introduction of automation, affecting over six million workers.

There are fewer formal mechanisms whereby workers can provide input to the process of factory automation in France and Italy. Yet some firms (especially those that are both producers and users of AMES) recognize the need for the support of labor unions in spreading automated manufacturing.

Technical Absorptive Capabilities

Swedish and German user firms work more closely with producers than French enterprises in the design, production, installation, and maintenance of AMES equipment. In this respect, the Swedish and German pattern of automation parallels the Japanese approach. Many of Sweden's firms in the manufacturing industries are characterized by a similar integration as is found in Japan. Volkswagen, one of the world's largest producers and users of robots, has shown no interest in commercializing its models.[8] This emphasis on inhouse production to meet its own needs is the same strategy pursued by Japanese firms in the early development of their robot industry.

French manufacturers have recently become aware of the usefulness of automating the production process. They are developing a structure for coordinating technical innovations in the automation domain. Demand for AMES equipment in France is expected to grow steadily and rapidly, perhaps as much as 30-50 percent per year. The relative absence of technical cooperation between producer and user firms will undergo change as demand for AMES equipment increases.

Swedish firms place a high priority on coordinating AMES equipment into the overall manufacturing system. This trend is sure to continue. For instance, CAD/CAM systems have so far been developed as subsystems controlling only limited portions of the production process. However, these systems are continuously being integrated into larger and larger systems to bring the full advantages of automation into play.

FOOTNOTES

[1]H. Selg and J. Carlsson, *Trends in the Development of Numerically Controlled Machine Tools and Industrial Robots in Sweden* (Stockholm: Computers and Electronics Commission of the Ministry of Industry, 1980).

[2]See table: "World Robot Population," *Financial Times* (London), 16 July, 1982, p. 3.

[3]*The Promotion of Robotics and CAD/CAM in Sweden* (Stockholm: Computers and Electronics Commission of the Ministry of Industry, 1981).

[4]H. Selg and J. Carlsson, "Swedish Industries Robot Experience," 1982, *op. cit.*

[5]An example of this move towards AMES is the small European batch manufacturer, Flymo, a British subsidiary of the Swedish Electrolux group (itself a robot maker). Flymo purchased 23 robots in a specific effort to increase productivity. Under the previous production system problems arose every time new products were introduced to meet variations in market demand. With the introduction of the robots, Flymo was able to introduce a modular system where one worker is responsible for all the assembly tasks for a particular product. Quality control is also now handled by an automatic inspection station which takes only 35-40 percent of the time previously needed. The new system allows Flymo to under-price many rivals. The payback on Flymo's robots was rapid—no more than a year, and another asset of the system has been its reliability. See "The Factory With No Workers," *Financial Times,* 14 July 1982.

[6]"The Factory With No Workers," *Financial Times,* 14 July 1982.

[7]D. Chamot and M. Dymmel, "Cooperation or Conflict: European Experiences with Technological Change at the Workplace," AFL-CIO (Washington, D.C., 1981).

[8]Organization for Economic Cooperation and Development, *The Impact of Automated Manufacturing Equipment on the Manufacturing Industries of Member Countries*, Paris, 1982.

PART III

COMPANY PROFILES FOR
SELECTED AMES PRODUCERS:
UNITED STATES, JAPAN AND EUROPE

CHAPTER VII
COMPANY PROFILES FOR SELECTED
AMES PRODUCERS: UNITED STATES,
JAPAN AND EUROPE

The selected company profiles presented here provide a microeconomic perspective of the robot producers in the United States, Japan and Western Europe. They provide insights into company policy and strategy, product lines, production, marketing, international joint ventures, linkages back to suppliers and forward to users.

These individual company profiles are based largely on visits to company sites and interviews with executives, supplemented by company and other documentation. They present an uptodate view of the remarkably dynamic policy and technology in the robot producing industry.

UNITED STATES

AUTOMATIX, INC.

Overview

Automatix is a private firm started up in 1980 largely with venture capital. Its nine principals bring to it extensive backgrounds in both hardware and software elements of AMES. Automatix has gained a 4 percent share of the U.S. AMES market (1982) primarily through sales of arc-welding robot systems, and a smaller number of vision systems. Automatix' strength lies in its ability to integrate the equipment of various manufacturers into innovative systems, and in providing extensive user support to a small customer base. Public sale of stock (March 1983) may provide capital to finance needed expansion of manufacturing and marketing capacity for the company's continued growth.

Current Position

Automatix' product line is concentrated in the areas of arc welding, assembly, and inspection, and includes the following:

1) Autovision I system: vision system for inspection, sorting, etc. Its price is $25-$30,000.

2)   Robovision I arc-welding system: combines the Hitachi arc-welding robot, a Lincoln Electric welding power supply and positioners and Automatix systems support. Its price is $85,000.

3) Autovision II: a faster and more sophisticated version of Autovision I with a factory-hardened controller, which uses Motorola 6800 microprocessor. Its base price is $30,000. Autovision 4 with Statistical Quality Control has been announced. The AV 4 will process up to 1200 parts per minute with a 250 x 404 pixel field.

4)   Robovision II: arc-welding system with Autovision II vision system, capable of interfacing with advanced sensors and with CAD/CAM systems.

5)   Cybervision System: assembly and advanced materials handling system using vision, in combination with an Electrolux pick-and-place robot.

Automatix has been primarily a systems integrator, utilizing purchased robots, cameras, welding equipment, etc. The company moved in August 1982 to new facilities in Billerica, Massachusetts, giving it increased manufacturing capacity and the ability to make internally a growing percentage of its products. Automatix recently introduced a new welding robot (the AID 600 with 5-axis capability) and has demonstrated a laser-guided seam tracking system for welding.

Automatix sales in 1980 (its first year of operation) were $0.4 million. Sales in 1981 amounted to $3.0 million, about 2 percent of the entire U.S. market. Sales for calendar year 1982 were estimated to be over $9 million, representing approximately 4.2 percent of the total U.S. market.

Although Automatix stresses its capability in advanced techniques such as vision, basic welding products account for the majority of its rapid growth. In mid-1982, for example, Automatix sold six arc-welding robots (worth about $0.5 million) to French automaker Peugeot.

The nine principals have extensive backgrounds in the AMES field. Philippe Villers, a Harvard graduate with a master's degree in mechanical engineering from MIT, was cofounder and senior Vice President of Computervision, a highly successful maker of CAD systems and interactive computer graphics software. Automatix' other principals include two others from Computervision, two from Data General Corporation, one from GE's robotics group, one from Unimation's California R&D center, one from the robotics/vision program at the National Bureau of Standards, and a former research manager at Sperry Research Center. The gathering of this impressive array of talent in a new start-up company accounts for much of Automatix' success, and reflects what is often an advantage for U.S. industry—the mobility of human expertise.

Corporate Strategy

Automatix continues to emphasize its capability as a turnkey systems integrator. Controllers are an Automatix specialty, and feature a programming language (RAIL) that is easy for operators to use, have built-in capabilities for automatic adaption to machine tool peculiarities, and have ready tie-in to main shop computers. Hence robot systems can be part of overall plant control.

Arc-welding systems will continue to be a major portion of Automatix' sales, since demand is projected to increase rapidly for this robot application. Growth in demand for vision systems will be strong in the medium term, as well, and Automatix is thus well positioned to capture a growing market share.

In April 1983 Automatix announced a major OEM agreement with Hirata Industrial Machineries Co., Ltd. of Tokyo, Japan. Under its terms Hirata will supply its family of SCARA assembly robots for Automatix to integrate into its turnkey robotic systems. Initially a series of three models with two to four computer controlled axes each will be used. These robots are of a recent design developed in Japan for simple assembly tasks including those involving many robots each performing a portion of more complex assembly tasks.

Automatix was originally financed largely by venture capital. The company raised over $10 million from banks, venture capitalists, Harvard University, and M.I.T. With capital presently at $31 million, successful sale of public stock, together with growing sales revenues, could provide the capital needed to increase manufacturing capacity and marketing strength.

GCA CORPORATION

Overview

GCA is a 24 year old company that grew during the 1970's to a level of $200 million in annual sales largely by supplying productivity-enhancing process equipment to the semiconductor manufacturing industry. Using capital generated by those operations, it entered the AMES field in three steps: by acquiring an established manufacturer of unique industrial robots (PaR Systems) in 1981; by signing an exclusive licensing agreement to distribute and service a line of general purpose industrial robots from Japan (Dainichi Kiko) in 1982; by establishing a well-staffed Industrial Systems Group (ISG) to manage its marketing, manufacturing and R, D&E functions in the AMES field.

The Industrial Systems Group (ISG) has a relatively young capability for inhouse R&D and manufacture of robots. Yet it has strong software design capability and management which is oriented toward the larger market for total computer-integrated manufacturing. While the fortunes of the parent company are still largely tied to the currently depressed market for integrated circuits, GCA's size and financial strength should permit sustained future growth by the Industrial Systems Group.

Current Position

1. Product Line

GCA's AMES product is one of the broadest available in the United States today from a single source. It includes the Extended Reach (XR-6100) Series of robots designed by GCA/PAK Systems, and the complete line of robots manufactured by Dainichi Kiko in Japan. GCA also offers a full line of robot controls adapted to each type of robot, ranging from the manual leadthrough type to sophisticated off-line programming with computer-simulated animation.

The XR Series robots capitalize on PaR Systems' experience in building remote manipulators for handling toxic substances, particularly nuclear materials. XR Series robots have up to six degrees of freedom and are of the overhead rail and trolley type, a configuration that is growing in popularity. They have a very large work envelope and high load capacity, up to 2,500 lbs. The sophisticated and relatively expensive ($150,000 per unit) XR Series robots are designed for applications in material transfer, machine loading and unloading, welding, and general assembly.

The Dainichi Kiko (DK) robots are a microprocessor-controlled series designed for general purpose industrial uses ranging from assembly to material handling applications. They range from a simple flat-plane robot with three degrees of freedom (F Series) to the multipurpose robot for heavy manufacturing (B Series) with six degrees of freedom. Payloads range from 11 pounds to 770 pounds for the moderately priced DK series robots.

2. Sales Volume and Market Share

Net sales of GCA Corporation were $218.5 million in 1981, while net income was $21.9 million. This reflects the strength of GCA's primary business, the manufacture of semiconductor production equipment, which accounted for 77 percent of sales and 84 percent of profits in 1981. Sales by PaR Systems, GCA's primary robotics division, are estimated at $19 million or about 8.5 percent of total sales in 1981.

3. Previous Experience Relevant to AMES

GCA's overall corporate goal is the manufacture and sale of industrial productivity systems. Its dominant position in the market for integrated circuit production equipment in the 1970's provided the capital and incentive to expand into the area of automated manufacturing equipment and systems closely related to the overall corporate focus.

GCA acquired, in June 1981, (for approximately $38 million) PaR Systems of St. Paul, Minnesota. Since 1961, PaR has designed and manufactured remote handling and large-scale programmable robotic systems.

To complement its newly acquired robotics manufacturing know-how, GCA hired Dennis Wisnosky as Group Vice President of the Industrial Systems Group (ISG) in October 1981. Wisnosky previously headed the factory automation program at International Harvester and earlier established the U.S. Air Force Integrated Computer Aided Manufacturing (ICAM) program.

Corporate Strategy and Structure

1. Structure for AMES

The GCA Industrial Systems Group includes PaR Systems; Vacuum Industries (makers of vacuum furnaces); and two new divisions: the Integrated Factory Controls group and the Integrated Manufacturing Systems group. These four divisions correspond to the structure and individual marketing functions for the four divisions within ISG. GCA/Vacuum Industries supplies equipment at the process level, PaR Systems manufactures robots and manipulators at the station level, Integrated Factory Controls will supply control architecture at the cell level and Integrated Manufacturing Systems will develop control software at the center level.

Collectively, the Industrial Systems Group can build the unified, modular "factory of the future."

The new GCA/Integrated Factory Controls division will develop the control architecture to actually manage machine tools and robots on the factory floor. Initially this division will focus on programmable controls through direct numerical control. Eventually, hardware and software systems capable of supporting the entire factory management will be developed. The Controls group will also study sensor technologies, such as vision, force-torque and ultrasonics, as well as traditional position and count status information.

The new Integrated Manufacturing Systems division will develop control systems for managing the entire manufacturing process in order to ensure smooth and continuous transitions from the strategic plans of management to tactical operations. The responsibility of the new Systems Group will lie in the area of soft technologies, i.e., software developments such as manufacturing resources planning and control, group technology, computer-aided process planning, simulation, etc.

2. Strategy for AMES

GCA's strategy to secure an immediate market presence has been to secure an exclusive distributor and licensing agreement with Dainichi Kiko of Japan for GCA to market, sell, service and support Dainichi Kiko's line of industrial robots in the United States and Canada. GCA thus gains immediate benefits from DK's ten years of experience in robotics, its broad product line, and its acknowledged capabilities in design, engineering and manufacturing of industrial robots and controllers.

While DK robots will initially be imported, the agreement foresees their production by GCA in the United States, using imported production technology. With its production capacity and product line established, GCA can devote more of its efforts to research, design, and engineering of new products, and to marketing.

GCA and DK reached a basic accord in May 1982 to conduct an extensive joint technological venture to develop new types of intelligent robots and CAD/CAM systems. Under the new agreement, the two corporations, besides developing new "seeing" and "feeling" robots and CAD and CAM systems, will enlist the cooperation of various American venture business enterprises, academic and other research institutions in their joint projects, through commission services or direct participations.

GCA's Industrial Systems Group (ISG) has focused its marketing and sales effort on these industries:
—Automotive and Automotive Support
—Foundry and Forge Operations
—Aircraft and Aerospace, Primarily Government and Military
—Heavy Equipment Including Trucks, Off-Road Vehicles and Locomotives
—Oilfield and Energy Equipment
—Electrical/Electronics Component Assembly
—Light Manufacturing, Particularly Appliances
While ISG is prepared to sell individual robots (as capital equipment), it is targeting customers interested in total computer-integrated manufacturing. ISG could thus supply everything from initial consulting services through robots and controllers to software.

GENERAL ELECTRIC CORPORATION

Overview

GE is a large diversified manufacturer which has developed considerable expertise in robotics through an ongoing program begun in the mid-1970's to automate its own plants. Since 1980, GE has spent over $500 million on acquisitions, joint ventures and licensed technology to become a supplier of complete (hardware and software) factory automation systems. GE has the benefit of its existing sales, service and international marketing organizations in its AMES effort. It will stress modularity, flexibility, and the ability to interface its equipment with that of others in an effort to capture 30 percent of the U.S. market by 1990.

Current Position

By licensing technology from Japan's Hitachi Ltd., Italy's OEA SpA, and West Germany's Volkswagen, GE has transformed itself from knowledgeable robot user to broad-line robot supplier. GE's "Allegro" multi-arm, light assembly robot system, Pragma, was introduced in April 1981. Since November 1982, GE has been selling five different robots licensed from Hitachi. These include a process robot, two seam-tracking arc-welding robots, and two paint-spraying models (vertical and horizontal configurations). In mid-1982 GE began to sell, under license, five heavy duty robots built by Volkswagen. These include a spot-welding robot, one for palletizing/stacking, two general purpose process robots, and a materials handling robot.

GE's line of eleven industrial robots is complemented by several other products. Its "Optomation II" is a microprocessor-based robot vision system capable of handling four cameras that sells for about $42,000. It has been demonstrated in combination with GE's Allegro assembly robot. GE also produces numerical controllers for machine tools, and microprocessor-based programmable controllers designed to interface not only with its own robots, but with those of other producers. Estimated external (outside GE) robot sales for 1982 are $1.5-2 million, giving GE a market share of about 1 percent.

GE recently announced an advanced vision and control system which enables a welding robot to steer itself along irregularly shaped joints, continually observing the joint and weld "puddle" and making adjustments as it travels along.

The promised payoff, according to a GE spokesman, will be a 15-fold increase in productivity compared with conventional welding robots, along with weld quality equaling or surpassing that of an expert human welder.

The robot vision and control system was invented by engineers at General Electric's Schenectady (N.Y.)-based Research and Development Center, which has applied for several patents on the technology. The system's vision sensor is based in part on a concept of viewing the weld puddle along the electrode first demonstrated at Ohio State University's Center for Welding Research, a university/industry R&D organization supported by GE and others.

Extracting weld puddle information from the scene observed by the vision sensor is a key capability of the system. It continuously analyzes the weld puddle geometry and feeds this information to an "adaptive welding process controller" that determines the corrections in welding conditions necessary to maintain the desired puddle geometry. The arc current corrections are then transmitted to a controller on the robot's welding power supply that automatically changes the current output.

GE operates a "rent-a-robot" program that furnishes units to its various plants on a try-out basis. Requests for such robot rentals often orginate with manufacturing management in individual GE plants. Operations managers take a course aimed at more fully integrating manufacturing operations with GE's overall strategic plan, and to provide tools and techniques to accelerate productivity improvement in manufacturing. Each course participant then undertakes some new program—directly spawned by his or her course experience—to realize significant productivity improvement. GE's top-down commitment to AMES is thus complemented by bottom-up interest and action.

GE's latest labor contract (with the United Electrical Radio, and Machinery Workers—UEW) broke new ground in easing the job displacement effects of factory automation. Creative solutions to this problem will be required for some time to come, since GE's plans call for up to 50 percent of its total blue-collar workforce of 37,000 to be replaced by some 14,000 robots through the 1980's.

Corporate Strategy

GE recognized the need to robotize itself to meet cost and quality goals in order to achieve the 6 percent annual productivity growth needed to remain internationally competitive. Internal experience with adoption of AMES since the mid-1970's has resulted in a pool of expertise in applications engineering within GE. Strategic plans call for up to two-thirds of GE's products to be "intelligent" (through incorporation of microelectronics) by 1990. Thus, three elements encouraged GE to enter the market as an AMES producer in 1981: factory automation experience inhouse; its position as a diversified manufacturer; and growing emphasis on microelectronics.

GE's overall strategy in the field of AMES is to become a leading integrator of factory automation by providing all the required elements. Since 1980 GE has invested $500 million and intends to spend another $250 million in assembling the building blocks to become a dominant producer of AMES.

GE has divided its effort in robotics into five areas to be worked on by interdisciplinary teams from corporate R&D and individual GE divisions: mechanical structure, robot drives, controls, sensors, and software. In all of these areas, GE emphasizes modularity, flexibility, and interfacing capabilities in order to achieve the widest possible applicability for its products.

Two recent acquisitions and a joint venture give GE significant software capability in the CAD/CAM/CAE area. GE spent (in 1981) $170 million to acquire Calma Corporation, a strong software house already producing CAD systems. GE also acquired (for $235 million) Intersil, Inc., a leader in complementary metal oxide semiconductor technology involving integrated circuits that can withstand heat and electrical distortion on the factory floor.

In early 1982, GE formed (spending $10 million) a joint venture with Structural Dynamics Research Corporation, whose software allows engineers to model parts and perform stress, vibration and other analyses on them. This gives GE an entry into the market for CAE systems as well as providing it with powerful software for its own efforts. GE is opening several productivity centers which will enable customer companies to use the software. Present centers are located at Cincinnati and San Diego. Others will be opened in Tokyo, Detroit, Wiesbaden and London.

GE signed a five-year technology exchange agreement with Volkswagen in early 1982. GE sells five large VW robots here under license, and may eventually produce them here as well. GE sells (and may produce) the "Allegro" robot under license from DEA of Italy. The flow in this case is one way, as the agreement does not call for technology sharing. GE's licensing agreement with Hitachi provides for selling Hitachi robots and for technical exchange.

GE has made its Optoelectronics Systems Division (maker of robot vision systems in Schenectady, NY) part of the Automation Systems Division (headquartered in Bridgeport, CT), which has overall control of the AMES effort. GE recently completed a $37 million complex (near Charlottesville, VA) to house manufacturing facilities and the headquarters of its Industrial Electronics Group. Itself highly automated, this complex produces electronic controllers and houses a new industrial automation R&D lab.

GE, like other AMES producers, is confronted with a shortage of engineers with multidisciplinary talents. Yet it has an advantage in building its AMES business on service, support, and applications engineering of licensed technology, since a core of expertise has been built up through internal robotization. GE's existing industrial sales force will be used to sell robots. The sales force is comprised of engineers who already have customer contacts and experience with highly technical equipment. They will simply be trained to handle a new product. Also the Automation Systems Organization will use GE's existing service organization, which is accustomed to servicing complex industrial equipment and will simply add robots to its range of competence.

GE's newly-formed worldwide marketing organization, General Electric Trading Company, will be used to sell GE's AMES products (as well as those of client manufacturing firms) abroad. The company will rely heavily on GE's in-place international marketing structure, including sales offices in 55 countries.

Finally, a new high-level communications language (to permit interchange of different robot programming languages) is being developed at Intersil. GE is working on a low-level factory communications system (called GEnet) built around its programmable controllers. These two communications links may allow GE to integrate a variety of automated manufacturing equipment (NC machines, robots, CAD systems) from different manufacturers into one computer controlled system.

GENERAL MOTORS CORPORATION

Overview

General Motors, primarily a vehicle manufacturer, is the largest user of AMES in the United States. In order to remain competitive in the vehicle field, it has planned extensive automation of many production, design and engineering functions. As part of this plan, GM entered a joint venture with a strong Japanese AMES producer, Fujitsu Fanuc. The resulting firm, GM Fanuc Robotics (GMF) holds the potential to be a major competitor in the AMES field, given the complementary strengths of the partners.

Current Position

As a major producer of vehicles and transportation-related equipment, GM was one of the first U.S. firms to utilize robots in production. The Puma robot, now built and sold by Unimation, was originally conceived and specified by GM to handle expected growth in parts assembly, machine loading, and parts handling. GM's leading robot suppliers at present are Unimation, Cincinnati Milacron, Copperweld Robotics, and Prab Robots Inc.

Auto sales have declined or remained sluggish since the late 1970's, while international competition (primarily in terms of price and quality) has increased. These factors led GM to undertake an extensive plan to automate many production, design and engineering functions over the next decade.

GMF may market a numerically controlled robot painting system developed by GM and tailored to its own needs. GM is also among the leaders in machine vision development and application, evolving from Sight 1, GM's first vision application in 1977 for IC chip alignment. GM now has 60 vision systems used in production at that original location. Another vision development is GM's Consight system.

Corporate Strategy

1. Robot Usage

In spite of being already the single-largest consumer of robots in this country, GM will accelerate its purchasing of robots from a cumulative level of 1,600 in 1982 to an estimated 14,000 by 1990.

Spot-welding robot operations account for 43 percent of all GM robot installations thus far. GM expects to have some 14,000 installed robots by 1990, including welding (19 percent), parts assembly (36 percent), machine loading (29 percent), painting (16 percent), and transfer and inspections (16 percent).

2. Corporate Structure for AMES

In late 1980, GM formed the GM Corporate Robotics Council to coordinate the rapid growth of industrial robots throughout GM. GM has also established a corporate center for testing and evaluation of robots.

GM's robot lab has been visited by several thousand of GM's engineers and managers. This is one method of providing to company staff a broad exposure to AMES equipment and concepts, and of increasing their acceptance.

GM also provides notification and discussion with local unions prior to introduction of new technology and provides short-range, specialized training programs to assist employees in adapting to new or changed work assignments.

The General Motors and Fujitsu Fanuc joint venture (GMF) is to design, manufacture, and sell robots. It is to be owned equally by the two partners and headquartered in Troy, Michigan. This represents a significant change for GM which has not entered into joint ventures since 1940.

Eric Mittelstadt (formerly executive assistant to the vice president for planning of GM) will be the president and CEO of GMF; and Seieumon Inaba (president of Fanuc) will serve as chairman of GMF. The agreement calls for eventual production of AMES in the United States, but a site has not yet been announced. GMF, with 50 percent ownership by each of the partners, will be operated as a separate entity from GM, though there will be some 25 GM executives on its board.

Any stepup in GM's robot purchasing may ultimately benefit the overall industry and certain individual companies in that it would serve as a rather significant motivation to other potential end users (particularly GM's competitors). Although orders of robots by GM will still be forthcoming to outside vendors, it stands to reason that certain companies will be more negatively impacted over time than others. Current suppliers will not be entirely stalled in their sale of AMES to GM, although the number of suppliers will likely decrease, and those remaining will be judged against GMF's capabilities and standards.

Whether GMF makes a major effort to market AMES outside of GM or not, General Motors Corporation obtains immediate benefits from the joint venture, including access to Fanuc's strength in design and production engineering for AMES, its status as a large, low-cost producer of AMES, and its good reputation in the AMES field. The underlying motivating factor for GM, however, is the fact that jointly produced robots can be *readily available*, without the three-to-five-year lag needed for development of a comparable line of robots by GM alone. This is the same motivation at work in GM's recent announcement of plans to produce small cars in cooperation with Toyota at its vacant plant in Fremont, California.

GM's extensive knowledge of the U.S. AMES user environment and experience in robot applications are attractive to Fanuc, which also gains access to the U.S. market through GM's marketing and distribution system, financial strength, and reputation. GM's myriad manufacturing operations in the United States and abroad also provide an excellent "test bed" for R&D on new AMES by GMF.

The combination of assets of both partners makes GMF Robotics an important actor in the commercialization of AMES, and a potentially strong competitor.

PRAB ROBOTS, INC.

Overview

Prab Robots, Inc. is a medium-sized firm that emphasizes simple robots, sold to a broad range of users, in small quantities. It subcontracts production of a large proportion of its AMES products, emphasizing long-term, quality-oriented relationships with component suppliers. Prab acquired a new manufacturing facility in 1981.

An early start (1969) and broad experience in user applications have given Prab a relatively large share of the U.S. AMES market (6 percent in 1981). Prab has pursued foreign sales through licensees in Japan, Canada and Europe.

Current Position

Prab grew out of a company started in 1961 by its current president and chairman, John Wallace. That company was Prab Conveyors, Inc. which began developing industrial robots in the late 1960's. Pick-and-place machines were first sold for die-casting operations in 1969.

In 1979, Prab purchased the Versatran robot line from AMF, Inc. These are relatively sophisticated, microprocessor controlled robots, often rail-mounted, with capacities of up to 2,000 pounds with a price range from

$45,000 to $150,000. "Prab" robots, developed internally, are simpler electro-hydraulic models with capacities up to 125 pounds and priced from $25,000 to $40,000. In addition to its eight robot models, Prab has developed a new controller, which will enable the company to sell its E and F models for $10,000 less per robot, and improve startup time by a factor of ten.

Prab supplies robots to users in the machine tool, plastics, glass, foundry, appliance, automotive, chemical, and electronics industries. Of over 1,000 Prab robots installed, including Versatran robots installed by AMF, only 50 are in automotive spot welding. In areas such as material handling, machinery, etc. (versus automotive spot welding), 90 percent of Prab's robot shipments are retrofits to existing capital equipment.

Corporate Strategy

Prab's corporate slogan is:"keep it simple." Although its product line is relatively broad, it emphasizes that users should not buy an "overqualified" robot, since the present level of technology seems to be more than U.S. industry can absorb. Research and development on robotics thus receives a relatively smaller share of Prab's capital budget. Prab has been involved, from the beginning, with smaller companies, and 60-80 percent of its business is with companies not in the automotive sector. In rough order of importance, it has targeted applications in palletizing, injection molding, die casting, material handling, parts transfer, machine tool load/unload, forging, casting, spot welding and induction heating.

Prab's emphasis on smaller orders (and non-pursuit of large, auto industry spot-welding orders) is partly the result of a production capacity constraint. Prab subcontracts a substantial amount of its robot components to nearby firms. The entire production of Prab non-servo robots and other robot components is subcontracted to Robot Research Corp. of Michigan (run by one of the designers of the original Prab robot), in a long-term relationship that continues to ensure high quality and low cost.

Prab prefers to concentrate on a marketing program involving a large number of customers ordering small numbers of units. This policy has helped the company develop a wide-ranging applications expertise, and a work force consisting of approximately one-third direct factory labor and two-thirds management, technical, and sales personnel. Its broad market base insulated Prab from some of the shocks that other AMES suppliers experienced when the recession hit the auto and appliance industries, two of the largest users of robots.

Every buyer of a Prab robot receives training for two employees at Prab's plant. The importance of this goes beyond the need for competent robot operators. Prab stresses the need to overcome the resistance of middle management, and the development of a broad base of trained personnel (in operations, applications, and service) in user industries.

Prab has also emphasized foreign sales of robots, especially in Japan and Europe. Industrial robots are sold outside the United States by distributors in Australia and New Zealand, and in South America by an agent. In September 1981, Prab entered into a license and technical exchange agreement with Murata Machinery, Ltd. granting to that firm an exclusive license to manufacture and sell Prab robots in Japan and other Far East countries. Prab also has a licensing agreement with Can-Eng Manufacturing, Ltd. of Niagara Falls, Ontario for the manufacture and sale of Prab robots in Canada.

In March 1982, Prab licensed Fabrique Nationale in Belgium to manufacture and sell Prab robots throughout Western Europe, Africa, and Scandinavia. The tie-up with Murata is especially significant, since Murata is well established in the FMS field. Prab's robots will complement Murata's systems effort, while the technological exchange agreement will enhance Prab's ability to supply systems-oriented customers in the U.S. Prab's guideline in taking on robot systems work is that 60 percent of it must be of Prab content.

UNIMATION, INC.
Overview

Unimation, Inc. is the pioneering U.S. robotics firm, begun in 1961. It achieved a 40 percent share of the U.S. market during 1980-81. Unimation has been a leader in R&D. Its emphasis on large, sophisticated robots has lately been broadened by the addition of smaller, lighter models and vision systems. Its business continues to be dominated by automotive sector buyers—a negative factor during the current recession.

Current Position

Unimation offers a broad spectrum of robots grouped into three major lines: its Unimate, Apprentice, and Puma robots. Between 90 and 100 robots are produced each month. Unimate's licensee, Kawasaki, the second largest robot manufacturer in the world, ships 60 Unimates a month.

Unimate robots are heavy-duty, hydraulically-powered machines with relatively great arm extension and weight-handling capability that have been widely used in spot-welding and machine load/unload applications. The Puma, developed in cooperation with General Motors and introduced in 1979, has an electric motor-driven arm and computerized memory that enables it to perform intricate motions particularly suited to a variety of assembly functions. The Apprentice, developed for a Finnish shipping company, is relatively small and easily movable and is designed for on-site arc welding in confined spaces.

"Univision I" is Unimation's vision system for its Puma robots. Machine Intelligence Corporation supplies vision-processing components and software for the system.

The automotive sector purchases the major portion of Unimation's output. The company estimates that over 60 percent of its FY 1982 shipments and 65 percent of its current order backlog are related to the automotive industry. Thus, Unimation has suffered more than any other producer from the auto industry's cutback in capital spending during the current recession.

Unimation was created in 1961 by Consolidated Diesel (Condec) to exploit a package of robot patents it had purchased. The impetus behind Condec's move was provided by one of its employees, Joseph Engelberger, who later became president of Unimation. Always a leader in robotics innovation, Unimation only became profitable in 1975. However, it has accounted for an increasing share of Condec's sales since the U.S. robotics market began to grow rapidly in 1979.*

Corporate Strategy

Unimation's product emphasis has been in the large, sophisticated industrial robots that can be easily reprogrammed to perform a wide variety of tasks. A Key ingredient to future growth in the AMES industry is a firm's ability to offer state-of-the-art technology. Recognizing this fact, Unimation is making a strong commitment to research and development. The company has more than 90 engineers working on vision, tactile sensing, multiple appendages and mobility, and, of utmost importance, enhanced software in the form of easy-to-use programming languages. Unimation West, its R&D laboratory in California, is perhaps the most advanced in the U.S.

Unimation opened a new plant in late 1982 that should provide the firm with added vertical integration in the manufacturing process. Production costs should be further reduced by the consolidation of existing facilities to create a more efficient manufacturing operation.

Kawasaki Licensing Agreement

Unimation licensed Kawasaki Heavy Industries in 1968 to manufacture Unimation robots in Japan and the Orient. Unimation has the right to receive, without charge, any technological enhancement developed by Kawasaki. Unimation will be offering a paint-spraying robot that was developed by Kawasaki and successfully introduced in Japan. As a result of a joint development effort with Kawasaki, Unimation has introduced a new robot (Puma 760) designed for mechanical assembly and arc welding, and is now working on a new mechanical configuration for press transfer application, as well as a more advanced vision system for Unimation's arc-welding robot.

---

*As this book was going to press, acquisition of Unimation by Western Union was announced.

Unimation has derived considerable benefit from fees and royalties (as well as name recognition) through this agreement. To date, the flow of technology has been primarily from Unimation to Kawasaki, but this is changing. Ford Motor Company has recently expressed strong interest in some of Kawasaki's 6000 Series Unimates. These robots, which have several articulated arms and a control system that can accommodate 48 axes of motion simultaneously, are the result of Kawasaki's R&D program. Unimation, in this case, would act only as U.S. supplier to Ford.

European Subsidiary

Unimation sells robots in Europe through its sales and service office in West Germany. The company also has a manufacturing subsidiary in Telford, England, which produces around 15 Puma's per month using 96 percent United Kingdom-sourced components. Fifty percent of this plant's output is sold in the United Kingdom while the rest is exported to Europe.

Union Carbide Agreement

In April 1981 Unimation signed a joint agreement with the Linde division of Union Carbide Corporation (one of the top four welding equipment makers in the United States) to market robot systems for automatic arc welding. Systems will offer state-of-the-art technology combining Linde DIGIMIG microprocessor-controlled welding equipment and Unimation's Puma and Apprentice robots. The two companies are also collaborating in producing gantry-mounted robot systems. Linde has over 300 U.S. distributors who will sell the robots, and Unimation will help work out turnkey installations.

Financial Capability

In October 1981, Condec Corporation, Unimation's parent company, announced that it would sell about 20 percent of Unimation to the public, thereby enhancing Unimation's financing capabilities and enabling Condec to rid itself of a portion of its heavy debt load. In the initial public sale of Unimation stock, 1.1 million common shares were sold at $23 each. Net proceeds were used primarily to repay some $19.4 million of outstanding long-term advances from Condec. In addition, a $10 million credit line was arranged with a group of six banks in November, 1981. The company did not anticipate borrowing significant amounts under the arrangement during fiscal 1982.

Unimation's net income has increased steadily from $880,000 in 1977 to $1.95 million in 1981. Since the end of FY 1981, the company's backlog has been reduced from $46 million to $35 million as of March 1982. It is expected that after this backlog is shipped by the end of fiscal 1982, Unimation will enter fiscal 1983 with about $20-25 million in new orders. Yet firm orders have declined in 1982 to about 45 per month.

In addition to an unpredictable order flow, the company will be under financial pressure as it moves part of its operation into new facilities. Marketing and R&D expenses are also expected to rise. The estimated 1982-1983 price/earnings multiple of 29 for Unimation stock is a major negative because it can discourage investors, especially in vulnerable current market conditions. Unimation has little long-term debt, and if it can survive the current recession, should be in good financial shape for a period of growth.*

JAPAN

AIDA ENGINEERING

Overview

Aida Engineering remains a leader in robots for press working operations. Aida's corporate strategy in the robotics industry continues to emphasize high-level technology, an international orientation, and financial conservatism reflected in a low debt-to-equity ratio by Japanese standards.

Current Position

The smallest firm among the leaders in Japanese robot sales is Aida Engineering, Ltd. In 1980 the company reported robot sales of 1.1 billion yen, the sixth largest of any Japanese firm. This ranked Aida ahead of such prominent robot producers as Fujitsu Fanuc (.8 billion yen) and Shin Meiwa (1.0 billion yen) and not far behind the 1.5 billion yen in robot sales reported by Hitachi.

Aida's relative position in overall robot sales fell somewhat in 1981 to the twelfth position, but it remains a leader in robots for press working operations. Robot sales in 1981 amounted to 1.6 billion yen, some 5.1 percent of total sales. In 1981, only two firms reported a higher dependence on robot sales than Aida: Yaskawa (7.1 percent) and Osaka Transformer (6.3 percent).

Aida is particularly strong in robots for press working operations, controlling 40 percent of the market. The company is now marketing several types of robots used in the manufacture of components for the motorcycle, bicycle, and automotive industry.

---

*Addresses for profiled companies are listed in *RIA Worldwide Robotics Survey and Directory* (Dearborn, Michigan: Robot Institute of America, 1981).

Corporate Structure and Strategy

Aida emphasizes high-level technology, an international orientation, and financial conservatism.

Industrial robots fit into the high technology image Aida seeks to promote. Given the danger of work involving the loading and unloading of large presses, the use of robots for these tasks seems highly desirable. Aida introduced the "Auto-hand," a machine the company considers to be an industrial robot. It has since actively continued the development of its robots technology, particularly technology that fits well with its major product, hydraulic presses. Aida has developed a series of loading and unloading robots for use with its presses and transfer lines.

In developing new technology, Aida has occasionally participated in government-sponsored projects. Aida was, for example, one of twenty firms participating in the seven-year national research project on flexible manufacturing systems organized by the Ministry of International Trade and Industry (MITI). This program involved three research institutes from the Ministry as well as the twenty manufacturers of materials, machine tools, and controls. Initiated in 1978, the project has a projected cost of 13 billion yen and is intended as a major step in making Japan the world leader in systems that can automatically produce parts in small lots in a wide variety of configurations. Aida's piece of the project is the development of a forging machine for small stepped shafts that would allow some changes in shape. This is to be a high precision machine that will include multiple forging positions.

Aida has a strong overseas orientation exemplified in several ways. Exports, notably to the U.S.S.R. and Southeast Asia, account for some 30 percent of sales. The company also has an American subsidiary, Aida U.S.A., which was established in December 1974. Aida recently concluded a contract with West Germany's Estel Hoesch Werke AG to provide technological aid involving CAD and CAM systems for metal molds. Estel will also market the LCDM CAD/CAM system in West Germany through its affiliated machine parts maker, Estel Rothe Erde Schmiedag AG.

Another contributing factor to Aida's success is its financial conservatism. While most Japanese firms have a debt-equity of about 30 percent owned capital, Aida's ration is 50/50. Major stockholders in the company are its banks and other institutional investors. The Industrial Bank of Japan and Fuji Bank each holds some 6.6 percent of the stock, Daiichi Marine Fire Insurance another 5.9 percent and C. Itoh, the giant trading company, holds 4.6 percent. Foreign ownership amounts to less than 2 percent.

AMADA

Amada designs and manufactures CAM systems for machine tools; CAD systems are developed by the users. The present systems evolved from NC machines which were first built ten to twelve years ago. The technology was developed by a research and development group set up in the United States (Seattle, Washington). The Japanese team did detail engineering on the first CAM prototypes and also adapted control system technology from the Hughes Tool Co.

Amada had problems in adapting both the Seattle and Hughes prototypes to Japanese user demands. The problems were related to the equipment's sensitivity to the more humid Japanese climate and inadequate sturdiness. The Hughes system was eventually replaced with a FANOL system.

This core technical group (four Japanese engineers) was later renamed Amada U.S. and moved to Los Angeles. The new group includes people with a technical background in computer science and electronics; two were trained at the University of California, others in Tokyo and Waseda universities. The group is now working on a laser processing machine.

Amada is not participating in the MITI-coordinated project as is Yamazaki. A previous joint development project (on a plate processing machine tool) with Mitsubishi Heavy Industries under MITI sponsorship originally budgeted for 200 million yen (half to be funded by MITI), eventually required 600 million yen to complete.

The major problem for Amada has been detailing user needs and designing appropriate complete CAD/CAM systems. CAD software is a particular problem. Computer makers such as Hitachi are better placed to develop this integrated software.

In the robotics area, Amada now delivers about six FMS systems a year. Each unit has about 40 major machine and control elements, and each unit has to be tailored to specialized customer need; such tailored units are costly.

DAINICHI KIKO

Overview

Dainichi Kiko, a Japanese firm founded specifically for the manufacture of robots, is noted for its concentration on exports. This emphasis is likely to continue as Dainichi Kiko actively seeks venture and marketing partners in the United States and Europe. The firm has been very successful in doing so and recently concluded a series of marketing agreements with GCA of the United States.

Dainichi robot technology combines high quality with an added emphasis on comprehensive after-sales support and maintenance.

Current Position

Dainichi Kiko is a small company that started out by developing the 25 ton "Husky" robot. This robot was used in large-scale welding inside the Sasago Tunnel on the Chuo Expressway in 1976. In 1978 Dainichi Kiko developed a smaller Husky-type robot with a built-in microcomputer. In 1980 the company marketed a new series of playback robots.

Most of Dainichi Kiko's robots are programmed to perform numerous industrial tasks such as palletizing, handling, sealing, cutting, machine loading, welding and assembly. In general, Dainichi Kiko's robot product line is very sophisticated using microcomputer control systems they themselves design and fabricate.

A new factory completed in August 1982 should add 100 more robots per month to achieve a total productive capacity of around 150 robots per month.

Corporate Strategy and Structure

Dainichi Kiko is a relatively new firm established in 1971 specifically for the manufacture of robots. Because the company is not well known in Japan and has few ties with the various industrial groups, it has had difficulty selling its robots on the domestic market and therefore has concentrated on exports—foreigners, according to Dainichi Kiko, are more concerned with the quality of the product than with the domestic status of the manufacturer. Currently, Dainichi exports at least 60 percent of its products.

The company's emphasis on exports is reflected in a series of agreements with several foreign firms. A recent report says the company will design and build material handling robots for Cincinnati Milacron, the second largest U.S. robot producer. In June 1982, the company announced the marketing of a new series of welding robots that would be sold through GCA in the United States. GCA has agreed to start producing Dainichi Kiko's robots by importing the necessary technology. The two corporations will also develop new "seeing" and "feeling" robots and CAD/CAM systems, enlisting the cooperation of various American research and business enterprise groups.

Besides its recent agreement with GCA, Dainichi Kiko has also a joint venture, Dainichi Sykes Robotics, in the United Kingdom to sell robots in Western Europe. This agreement is significant in that it is the first time that a Japanese robot manufacturer has been willing to transfer robot technology to a British company. The Dainichi-Sykes product line is claimed to be the widest range available from any robot manufacturer; it consists of ten different microcomputer controlled robots with a lifting capacity from less than a kilogram to over two tons.

Dainichi Kiko is also seeking agents in other parts of Europe. This strategy is part of an effort not only to sell robots but entire automation systems. The company has a qualified team of engineers working to ensure effective integration of the robots into manufacturing processes. Dainichi Kiko also provides comprehensive after-sales technical support and maintenance.

In 1982 Dainichi started to produce its own robot controllers and other major components reducing from 90 percent of the subcontracted components of their robots to 60 percent. Each foreign partner, however, continues to contribute to a percentage of the joint research and development program. The company has ten technical licensing and exchange agreements.

FUJIKOSHI

(Nachi) Fujikoshi is Japan's third largest producer of spot-welding robots (following Kawasaki Heavy Industries and Mitsubishi Heavy Industries). In 1981 it produced 230 robots valued at 1.8 billion yen—1.1 percent of total sales. Fujikoshi was the first Japanese firm to produce arc-welding robots. Robot sales in 1981, all under the "Uniman" tradename, were composed of arc-welding robots (50 percent), spot-welding robots (30 percent) and spray-painting robots (20 percent). The company plans to increase its robot production to seventy per month by the spring of 1983. By the end of 1983 it hopes to export 100 units annually to the United States.

FUJITSU FANUC

Overview

Currently Fujitsu FANUC ranks third in the production of robots and is likely to maintain or improve its current position, given FANUC's heavy emphasis on research and development, its export orientation, and its dynamic corporate leadership. It is probable that FANUC will be a forerunner in the development of new robot models and continue to expand its world share in robot sales. FANUC has been able to utilize its strength in the computer-controlled machine tool market by integrating robots into their systems.

Current Position

In Japan, Fujitsu FANUC in 1981 had a production volume of 600 units, three times its 1980 output, following the opening of a new $38 million robot factory. This new plant utilizes robots and NC machine tools to produce other robots and computerized tools. By 1986 FANUC plans to produce nearly 5,000 robots annually with only 200 employees. The Fuji factory is the first unmanned factory utilizing a total flexible manufacturing system in the machine tool industry.

The sales of robots increased dramatically from 1981 to 1982 from 800 million yen to 2,000 million yen. Numerically controlled systems continue to represent nearly 90 percent of total sales.

Currently FANUC markets three basic robot models in support of machining and assembly. FANUC's particular strength in robots is combined NC and robot equipment, where the robot does the loading and unloading for one or more NC controlled machine tools.

Corporate Strategy and Structure

FANUC is now the world's largest maker of computer machine tool systems, and its aim is to progressively integrate robotics into these systems.

One reason for FANUC success in the robotics industry is its strong orientation toward technological growth and development. In 1979 some 200 of the 730 employees were in R&D, around 330 were in sales and service, and only 200 were in production. FANUC was an early user of semiconductors and microprocessors in its numerical and computer controlled machine tool systems.

Another factor contributing to the company's success in the sale of industrial robots has been its orientation toward export markets. This is traceable to the fact that FANUC is a subsidiary of one of Japan's leading computer manufacturers, Fujitsu Ltd. Fujitsu continues to hold 47 percent of FANUC's stock; Siemens of West Germany owns 5 percent of the company and Fuji Electric another 6 percent.

FANUC has entered into several international arrangements. General Motors and FANUC have agreed to set up a company in the United States capitalized at $10 million as a 50-50 joint venture. The plant will be set up in Troy, Michigan and will initially market handling and assembly robots used in association with FANUC's computer controlled machine tools. Eventually the plant will produce robots on its own. The advantage of the FANUC-GM joint venture is that it combines the low cost base (due to its highly automated facilities) with GM's application experience in using robots.

FANUC also has ties with Manuchin Automatic of France in the field of industrial robots and is a partner in a South Korean venture. A Taiwan company, Tatung Engineering Company, has been granted the exclusive right to sell Fujitsu FANUC robots in Taiwan over the next several years.

FANUC has agreed to join with Fuji Electric and Fujitsu (computers) to jointly produce and market integrated machinery centers and flexible manufacturing systems. Under the agreement 1) FANUC would handle sales; 2) Fuji Electric would take charge of monitoring devices including automated conveyor and warehouse systems and display devices; 3) Fujitsu would handle industrial robots and NC machine tools and would also be responsible for the computers controlling the entire system; and FANUC will establish a Systems Engineering Office within the Technology Research Institute being constructed at its Fuji factory.

HITACHI

Overview

Hitachi has a strong position in the manufacture of robotic equipment based upon its overall size and technical capability and its experience as a leading manufacturer of computer and related electronic control equipment. The firm has intensified its research efforts in robotics and completely automated assembly lines. In cooperation with General Electric, they planned to double their exports of robotics and related systems by the end of 1982.

Current Position

Hitachi is ranked fourth in number of units. Robot sales were valued at 5 billion yen in 1981 (about 480 units, as compared to 150 a year earlier.)

By 1976 Hitachi was using some 600 robots in its own operations—most of them in materials handling. The know-how gained here would later be incorporated in Hitachi robots.

In 1975 Hitachi introduced its first welding robot (Mr. Aros) with a micro-computer control. This was a large and expensive machine, so cheaper versions were developed and marketed in subsequent years.

In another area, Hitachi worked with Nihon Parkerizing to develop a spray robot. The Hitachi-Parker Spray Robot is said to require less work space and thus be highly suitable for use in the automobile industry.

Another important development was the marketing at the beginning of 1980 of the Hitachi Process Robot. This is a low-cost robot for assembly and welding designed to compete with Kawasaki's Puma. It is expected that this robot will be produced in increasing numbers in the future.

Corporate Strategy and Structure

Hitachi initiated its first study of industrial robots in 1966. The early emphasis was on object recognition, and research was carried out at its Central Research Laboratories. In 1971 Hitachi founded the Production Technology Institute and began work on industrial applications.

In the past year considerable attention has been given the greatly stepped up research efforts on robots at Hitachi. It is reported that some 500 key technological experts are working on an intelligent robot for assembly. Hitachi is said to hope by 1986 to have some two-thirds of its assembly tasks handled by robots. The company is also working on new generation welding robots with new sensory equipment. A number of engineers from the company's operating divisions are in residence at the Hitachi laboratories to facilitate the transfer of technology to the plants. Hitachi plans to introduce 1,000 robots, especially for the assembly of home electrical plants. Hitachi is hoping that within five years, 60 percent of all assembly processes company-wide will employ robots and the number of operatives will be reduced by close to 30 percent.

Among the sets of expertise that are reinforcing Hitachi's potential in robotics is a strong background in computers. Despite an aversion to relying on foreign technology, Hitachi signed relevant technical assistance agreements with CDA in 1961. More recently, it has undertaken the development and production of large computer systems—some in cooperation with Fujitsu. Hitachi has also received considerable attention recently for its success in producing 256K random access memory chips which may find application in Hitachi robots.

Basically there are two objectives in Hitachi's robot strategy: (1) early attainment of automation and removal of humans from flexible manufacturing systems in internal plants and factories; and (2) acceleration of the development of high-performance, low-cost robots, experimentation in their own plants, and expansion of their market share in this field.

In January 1982, Hitachi announced plans to double robot exports over the previous year. It has set up a wide range of international agreements aimed at expanding exports; the most publicized is one with General Electric. GE will make three kinds of high-performance robots under license from Hitachi and will sell Hitachi-made robots under its own trademark. Hitachi is also to deliver about 200 high-performance robots to General Electric for marketing in third countries under the General Electric brand. The agreement will last seven years.

GE is not the only American firm selling Hitachi robots; another is Automatix of Burlingon, Massachusetts. Nor are Hitachi's efforts restricted to the United States. It has agreements with firms in Great Britain, West Germany, and France, and is reportedly negotiating license rights for sale in the U.S.S.R., Sweden, Finland, Belgium, Switzerland, Austria and Taiwan.

KAWASAKI

Overview

While Kawasaki is currently Japan's largest producer of robots, the firm's position of leadership is being challenged by other firms. Kawasaki has a ten-year renewable agreement with Unimation (United States), under which it markets Unimation robots in the Far East and has rights to Unimation patents.

If Kawasaki is to develop an intelligent robot it must dramatically increase the number of electronic technologists in its research and development division. Much depends upon whether Unimation will continue to provide Kawasaki with the requisite new technology, particularly microcomputers, sensors, and robot languages.

Current Position

Kawasaki Heavy Industries (KHI) is Japan's largest producer of robots, both in terms of value and number of units. In 1980 KHI produced 470 robots valued at 5.4 billion yen. The number of robots produced increased to 700 in 1981. While KHI's relative position in the Japanese robotics industry may be slipping, it is edging up on Unimation, the world's largest producer. Indeed, it may have passed Unimation in number of robots sold in 1981.

KHI's involvement in the robotics industry goes back to 1967 when it signed a ten-year renewable agreement with Unimation—the pioneer American builder of robots. Under the agreement, Unimation gave KHI the rights to Unimation patents, trademarks, detailed drawings and tooling. KHI markets Unimation robots in the Far East, and also sells variants of Unimates that it has developed in other parts of the world.

KHI began marketing Kawasaki Unimate robots in 1970. Demand for the robots was poor at first, but in 1971 the company delivered nearly twenty Kawasaki Unimate spot-welding robots to a major automaker in Japan. Since then KHI robots have been widely used in the auto industry, including Toyota and Nissan. Toyota Motor announced its intention of purchasing 720 robots over a three-year period from Kawasaki; 220 were delivered in 1980.

Kawasaki has also developed a paint-spraying robot that has been successfully introduced in Japan. Unimation will soon be offering this robot in the U.S. market.

The recently introduced Puma 76 model robot, designed for mechanical assembly and arc welding, was the result of a joint development effort between Kawasaki and Unimation.

Kawasaki is working on a new mechanical configuration for press transfer applications; this involves two programmable arms located between presses in order to move parts rapidly from press to press. Unimation and Kawasaki are also working together to design a more advanced vision system for Unimation's arc-welding robot.

KHI has exported Unimate-type robots worldwide, some to the Soviet Union. Some twenty welding Unimates are in service at the Kama River truck plant in the U.S.S.R., and KHI Unimate W2030 is being used at a ZIL automobile plant. KHI has also marketed nearly one hundred of its 6000 series robots to General Motors in the United States.

It should be noted that KHI uses its robots in many of its own plants—given the broad diversity of the company, this can give it a remarkable range of experience in industrial applications. At this time, nearly 90 percent of the domestic Japanese market in spot-welding robots is owned by Kawasaki. Most of the rest are used for casting, loading and unloading machines, palletizing, heat processing, painting and arc welding.

Corporate Strategy and Structure

KHI's position of leadership in the Japanese industry is being challenged by other firms. The company is responding to the challenge by building a new $76 million plant exclusively for the production of the robots. The new plant will be in Kobe, not far from the Akashi Works. KHI hopes the plant will produce more than 2,800 robots annually by 1987.

Kawasaki's current robot strategy, in alliance with Unimation, is to develop an intelligent robot by 1985.

KOBE STEEL

Overview

Kobe Steel, although not a major robot producer, holds 40 percent of the Japanese market for painting robots. Kobe Steel's agreement to market Trullfa (a Norwegian maker of robots) painting robots in 1973 led to its own manufacturing of Trullfa robots in 1978. Kobe's corporate strategy emphasizes the use of borrowed technology to improve robot models.

Kobe Steel, Japan's fifth largest steel producer, has sold more than 1,000 robots so far, mostly to the auto industry. It has held about 40 percent of the Japanese market for painting robots. About 40 percent of the painting robots produced by Kobe have "flexi-arms," that is, free moving wrists. Flexi-arms allow the robot to paint the inside of vehicles and corners that cannot be reached with conventional robots. Most of the flexi-arm painting robots are sold to the automobile industry—current users include Honda's Sayama Plant and Nissan's Tochigi Plant. Flexi-arm robots cost about $100,000 per unit (around 25 million yen).

Corporate Structure and Strategy

Kobe has emphasized using borrowed technology to make various improvements in its robot models. For instance, after beginning its own production of Trallfa robots, Kobe's engineers improved the robot's operating speed, reduced the force needed to teach the robots and found ways to apply them to conveyor systems. The most important of these was the reduction of the "teaching force" from three kilograms to one or two kilograms. Kobe, however, is also increasing its robot production target and hoped to produce a total of 300 units for 1982 and 400 for 1983.

KOMATSU

Overview

Komatsu, Ltd. is a leader in developing new applications for robotics technology, albeit not a major producer. Komatsu's current research strategy is geared towards developing an intelligent robot for doing assembly work. Most of its robot production has been used internally.

Current Position

Komatsu has been very active in finding new applications for robotics technology. Examples include an arc-welding robot that responds to verbal commands, robots for use in underwater bulldozers, and robots for the moving and assembly of very heavy parts.

One of the first Komatsu robots to attract attention was the Recus. This robot appeared at the Japan Industrial Robots Exhibition in Tokyo in October 1979. According to a report in *The Industrial Robot*, the Recus was already being used in surveillance work as the Japanese prepared to build a bridge connecting two of the major Japanese islands. Recus was reportedly designed to operate in the ocean at depths of up to 500 meters and in currents of up to 6 knots.

Corporate Structure and Strategy

Komatsu hopes to achieve practical application of the intelligent robot in 1983. Komatsu launched robot development in 1976 but the recent emphasis on robot research and development stems in part from the decision of Komatsu President Rijoichi Kawai to phase out the construction machinery department of the company. To facilitate development of intelligent robots for assembly work, a consolidated research center with investments totalling 15 billion yen is being built. The center will emphasize the electronification of machinery.

MATSUSHITA

Overview

Although Matsushita Electric Industrial Company has specialized in manufacturing robots for inspecting parts, its current strategy is to develop completely automated plants. Matsushita's strength in the robotics industry is sensor technologies, and the company hopes to exploit this advantage in developing intelligent robots.

Current Position

In 1981, the number of robots manufactured for part inspection exceeded 2,400. Matsushita also manufactures robots for tightening bolts and nuts; 30 percent of these robots are sold to other companies. Matsushita's robot sales were estimated at 600 million yen in 1981.

Corporate Strategy and Structure

Matsushita's robot strategy is to develop completely unmanned industrial plants. The company has only recently entered the field of selling arc-welding robots to outside companies. The sales are being undertaken by its subsidiary, Matsushita Business Machines. The management believes that more than 100,000 robots will be in operation inside the company within 10 years.

Under the guidance of a Robot Council and Robot Committee, formed in 1981, 15 laboratories and 40 working groups are now functioning; they are aiming at sales of 10 trillion yen a decade from now. The present project will concentrate on the development of intelligent robots having a certain degree of judgmental capability. Matsushita has developed various sensor technologies over the years and this technology is directly applicable to the sensory functions of the robots, which is Matsushita's strength in the robotics industry.

MITSUBISHI ELECTRIC CORPORATION

Overview

Mitsubishi Electric Corporation has the potential to emerge as one of the world's leading robot producers. Its assets include the fact that the company is Japan's third largest integrated machine company and it is also a central member of the Mitsubishi *keiretsu*, Japan's biggest industrial, trading and banking group, where it can draw on vast financial resources, trading connections and technology. Mitsubishi Electric is also noted for its inhouse research.

Although Mitsubishi Electric sold no robots in 1980, in 1981 it sold some two hundred units valued at two billion yen. In early 1982 Mitsubishi Electric had reportedly expanded production to thirty robots per month, and was projecting growth to be 50 percent per year over the near term future.

Mitsubishi Electric introduced an arc-welding robot at the Welding Fair in Tokyo in the spring of 1981. The correspondent for *The Industrial Robot* characterized the design of the robot as resembling that of ASEA's robot. Mitsubishi Electric, however, is uniquely self sufficient in all the requirements for the new robots. It makes its own microprocessors, integrated circuits, CNC controllers dc servo motors and precision machinery. Exports of the new robot were to begin at the beginning of 1982. In March 1982, General Electric announced that it would market Mitsubishi Electric welding robots. Mitsubishi Electric is supplying an arc-welding robot to Westinghouse on an OEM (original equipment manufacturer) basis.

Mitsubishi has also produced other robots for inhouse use. One washes windows at some of the buildings owned by the Mitsubishi group, another is used for die-bonding at one of the Mitsubishi Electric plants.

MITSUBISHI HEAVY INDUSTRIES, LTD.

Overview

Mitsubishi Heavy Industries, Ltd. ranks fourth in value of sales in the Japanese robotics industry. The company markets mainly painting, spot-welding and arc-welding robots. Although most robot sales are internal, Mitsubishi is attempting to establish more international markets for its robots, most recently the United States. However, it is concerned about increased protectionism in the U.S.

Mitsubishi Heavy Industries' significance may go beyond its size, since it is one of the core firms in the Mitsubishi Group, Japan's most powerful industrial, trading and banking group.

In 1975 Mitsubishi Heavy Industries entered a business tie-up with Iwata Air Compressor Manufacturing Company to develop a painting robot because of complementarities in sales, testing, and engineering and servicing. In 1980, Mitsubishi Heavy Industries independently developed a spot-welding robot. Its initial efforts were characterized by *The Industrial Robot* as "something that looks decidedly like a copy of a Unimate 2000." Originally they were used only in the plants of Mitsubishi Motors (a subsidiary). By mid-1981 250 had been produced—nearly one third of which were in service at Mitsubishi Motors' Okazaki Works.

Mitsubishi Heavy Industries also plans to market an arc-welding robot. One impetus for this is the company's need for a robot able to handle heavy plates at its shipyards and heavy machinery plants.

Mitsubishi Heavy Industries has some international ties for marketing, particularly in Europe. Sciaky, a leading French welding specialty firm sells Mitsubishi Heavy Industries welding robots in Europe. In October 1981, MHI announced a new agreement whereby Austria's largest industrial conglomerate, Voest-Alpine, would produce and sell Mitsubishi Heavy Industries robots under license.

Corporate Structure and Strategy

Mitsubishi Heavy Industries believes that flexible manufacturing systems are used by Japanese companies to "stay on top"; U.S. firms adapt to "catch up." Cost efficiency, flexibility, and quality are the three basic reasons for developing flexible manufacturing systems. In a growth economy, Japanese firms are forced to robotics to stay competitive.

Governmental funding is important for reinforcing demand by small utility companies. Mitsubishi Heavy Industries strategy is to link up with foreign manufacturers for co-production and marketing-law royalties and to sell certain critical components. Mitsubishi has teamed up with firms like Voest-Alpine on heavy equipment production and has worked at gaining technical partnerships aimed at avoiding patent infringements and providing certain engineering, e.g., software design complementarities.

There is deep concern about the prospect of increased protectionism in the U.S. machine tool industry (action of Houdaille is a case in point). Contervailing forces are a) U.S. firms that want to modernize to survive (GM, Caterpillar, e.g.,) and have found the U.S. machine tool industry inadequately responsive to their retooling needs (both in terms of products offered and delayed delivery times); and b) several firms like IBM and Westinghouse are not positioning themselves in world markets and are using international partnerships to provide marketing, production, and engineering complementarities that will give them an early global market position.

NIPPON ELECTRIC

Overview

Nippon Electric's involvement in the robotics industry has been limited largely to the production of robots used in the die casting process. More recently, the firm has entered the intelligent robot market concentrating on producing precision assembly robots.

Current Position

Nippon Electric Company is a large company with activity in electronics, computers, and telecommunications equipment. Robot sales were 220 million yen for 1981 and were estimated at 250 yen for 1982.

Nippon's intelligent robot development was undertaken several years ago for internal application. "Precision-assembly robots" capable of performing assembly and inspection of electronic parts have been supplied to Nippon's production lines. Nippon believes there is a large potential demand for a precision-assembly robot capable of handling small parts such as electronic parts. Currently, one of the three machine types commercialized by Nippon Electric has two arms operating independently or in concert, very high positioning accuracy, and program language for robot application. The robot can be applied to a wide variety of tasks in addition to assembly work.

OSAKA TRANSFORMER

Osaka Transformer Company (OTC) is Japan's second largest producer of arc-welding robots, and the equal of Mitsubishi Heavy Industries and Kobe in overall robot sales in 1981. OTC started selling robots in 1980, specializing in arc-welding robots for sheets. In the summer of 1981 it introduced a sealant/adhesive spray robot and a cutting robot. With the completion of a new plant, robot sales have increased from 120 units (one billion yen) in 1980 to 420 units in 1982 (four billion yen).

Osaka Transformer provides complete design, layout, production planning and installation for about 40 percent of their customers.

It has derived electronic control technology from other Japanese firms such as Sanyo and Yaskawa. The firm is currently faced with patent infringement charges by U.S. firms who claim that Okaka technology was copied from their models. To date, Osaka has not exported its robots.

PENTEL

In August 1981 Pentel, a pen manufacturer, began marketing the "Puha" brand assembly robot. Pentel's managers claim that before marketing the Puha they used it inhouse and that the robot lowered pen production costs by 45 percent while lowering the lead time for bringing new products on line by 80 percent. The Puha being marketed externally is designed for small, high-precision jobs in electronics, automobile parts and precision instruments. It is a programmable robot with a central computer processor that sells in Japan for about $19,000. As of May 1982, only two of these robots had been sold in the United States, but Pentel hoped to introduce more sophisticated models that would increase sales.

Because of the relatively slow growth in sales for its traditional products, Pentel has been aggressively moving to diversify in high technology areas such as robotics.

SANKYO SEIKI

Sankyo Seiki is Japan's second largest producer of assembly robots, behind Fujitsu Fanuc. In its first sales year, 1981, it sold 300 robots valued at 1.5 billion yen. Robot sales accounted for 1.3 percent of total sales. By 1985 the company hopes to sell more than 1,000 robots per year.

Five years ago Sankyo Seiki started doing research on intelligent robots and began selling a model with a single controller for about $20,000 in April 1981. Sankyo Seiki received considerable attention in February 1982 when IBM announced that it would enter the robot market by selling a low-cost Sankyo Seiki machine, the Skilam. The Skilam is an assembly robot that was introduced in June 1981. The robot was to be marketed in the United States as the IBM 7535 System at about $28,000 per unit, and shipments were scheduled to begin in October 1982.

SHINMEIWA

Shinmweiwa Industries is Japan's third largest producer of arc-welding robots (after Yaskawa and Osaka Transformer). In 1981 it sold 200 robots valued at two billion yen—about 3 percent of total sales. Shinmeiwa became a part of the Hitachi group in 1960; Hitachi owns more than 30 percent of the company's stock.

Shinmeiwa conducted research and development on robotics from 1969 to 1972, then sold its first arc-welding robot in 1973. In the specialized area of arc-welding robots for welding heavy steel plates, Shinmeiwa is Japan's leader. In 1980, Shinmeiwa reportedly exported 22 arc robots, including 15 to the United States. By the spring of 1983, the company expects to increase its robot production capacity to forty units per month.

Shinmeiwa exports through several companies: Grundy in Great Britain, Cammecy in France, as well as Grundig and AGA in Europe.

TOKICO

Overview

Tokico is Japan's third largest producer of robots for spray painting, led by Kobe Steel and Mitsubishi Heavy Industries. In 1981 Tokico sold a total of 120 robots valued at 1.2 billion yen. Robot sales accounted for 1.4 percent of total sales. Tokico, like Shinmeiwa, is a member of the Hitachi group; Hitachi owns about one-sixth of its stock. Tokico has ties with Ransburg Japan.

In the fall of 1981, the company added three new spray-painting robots to its line to strengthen its position. Tokico can now produce about ten robots per month, but plans to double its production capacity by the spring of 1983, and further increase it to about 50 units per month in two or three years. One strength of the company is its ability to produce most of the parts needed for its robots.

TOSHIBA KIKAI COMPANY

Toshiba Kikai evolved after 1970 from a manufacturer of numerically controlled machines to a producer of machinery centers and flexible manufacturing lines (based on Mill-OKI-Matic and heavy Trekka prototypes). Toshiba now has one hundred design engineers working in this field. The company has had a 50-50 joint venture with Kearney-Tickka which is now 100 percent under its control.

Toshiba competes against other Japanese equipment suppliers to supply machining centers, each costing over one billion yen. It takes two to three years to design and deliver such a center. During that time, market conditions or customer requirements may change, all of which cause complex logistical problems in planning such markets. Each flexible manufacturing system has distinctive characteristics in terms of product mix and production volumes.

FMS was developed in response to a shortened product cycle. In Japan, firms seem more capable to articulate changing requirements and to provide engineering, and the maintenance necessary to adjust designs to particular needs and to install new systems rapidly and effectively.

Conglomerate companies, such as Toshiba, are better able to draw upon diversified industrial capabilities for such systems. Toshiba is now extensively introducing flexible manufacturing systems into its own facilities (electrical switches, medical equipment).

YAMAZAKI

Current Position

Yamazaki is a major Japanese NC and computer controlled machine tool manufacturer which has recently developed a flexible manufacturing system (FMS). Yamazaki has over $350 million annually in sales and about 70 percent of its business is export. It is still a privately held firm with a staff of about 1,600 plus an additional 400 workers in overseas operations.

Yamazaki's flexible manufacturing system costing twenty million dollars is capable of inspecting finished products, rejecting or reprocessing bad products, remembering and performing simultaneously 80 different kinds of jobs, controlling the main axis revolutions according to the character of each job and its materials, and automatic switching of wornout or damaged tools.

Yamazaki is building a new FMS plant in Florence, KY. The new Minokama plant will employ 200 people (a pre-FMS plant today would employ 1,500). It can respond readily to changes in product demand. The new FMS system reduces materials in the inventory cycle from 80 days down to less than one day.

Yamazaki is planning for a six-month design delivery cycle. Realizing that product life cycles have been shortening dramatically since the oil crisis of 1973. Yamazaki believes its competitive advantage lies in being able to design and deliver newer models of machine tools in 16 months as compared to 24 to 36 months in Europe and the United States.

Supplier industry however, will have to become more responsive and capital-utilization more efficient (Toyota buys 83 percent of its vehicle content).

Many small firms in rural areas of Japan are able to pay for new machine tools from increases in land values and Yamazaki provides financing to component suppliers or preferred firms. Only 10 percent of Yamazaki product content is purchased from suppliers. Yamazaki is now attempting to develop its own CAD/CAM systems. These systems had been obtained from FANUC until recently, but are no longer available from that source.

About 70 percent of Yamazaki output is exported, of which 40 percent goes to the United States, 30 percent to Europe, and 30 percent to the rest of the world. Extensive training for customers, however, is done at Yamazaki's main plant.

YASKAWA

Overview

Yaskawa, Japan's second largest robot producer in 1981, is planning to increase its robot sales substantially over the new few years. Given Yaskawa's significant export orientation and its dependence on robot sales, it is likely that Yaskawa will continue to see high technology as its major growth area.

Current Position

Yaskawa produced 685 robots in 1981 as compared to 360 in 1980. In both years, Yaskawa was more dependent on robots as a percent of total sales than other producers—4.7 percent in 1980 and 7.1 percent in 1981.

Yaskawa received orders of more than 100 units per month in the latter half of 1981. Additionally, progress in servo-motor technology allowed for the development of a new multi-jointed type of arc-welding robot, and Yaskawa now dominates 70 percent of the market in this field.

Yaskawa's involvement in the robotics industry goes back to 1968 when the company developed a fixed-sequence robot. In that year, Yaskawa projected its annual output in 1981 would be 650 units (actually it produced 685). Of these, fifty were to be exported to the United States and fifty to Europe. Yaskawa was the first Japanese robot maker to establish its own overseas sales network, in five western countries.

Most of Yaskawa's robots are applied to welding steel sheets. It developed its best known model, the "Motoman," in 1972. In a June 1981 report, Yaskawa claimed to have sold a total of 550 Motomen L-10 robots. Some 80 percent of the L-10s were being used in the automobile industry, especially by subcontractors making components for the major producers; another 10 percent were being used in the electrical equipment industry. Yaskawa arc-welding robots are used at several of the large Japanese automakers including Nissan (Datsun), Mitsubishi and Fuji (Subaru).

Because the L-10 was too large for some applications, scaled-down versions were also introduced. For example, the Motoman L-3 is a miniaturized L-10. The L-3 is supposed to offer twice the accuracy of most robots, and weigh only 93 kilograms. It is about 15 percent cheaper than the L-10.

Yaskawa has recently expanded production of its sealant/adhesive spray robot, which it had developed. It also announced that it will start marketing an arc-welding robot using an arc-sensor system to measure the acting position of the unit. This was to come on the market in April 1982. The smaller Motomen and other new robot lines, notwithstanding, Yaskawa expects the L-10 to be its mainstay in the industry for the next few years.

Corporate Strategy and Structure

The company hopes to sell 50% of its robot output overseas by 1985. Due to its export orientation Yaskawa has extensive ties overseas with distributors. In the United States its products are handled by Hobart Brothers, Inc., a welding systems supplier in Troy, Ohio. In Great Britain it is represented by GKN Lincoln. In a recent joint venture, Yaskawa was linked with the Machine Intelligence Corporation, a specialist in robot vision systems. Yaskawa does very little overseas licensing because of fears that this would limit its markets. There is concern at Yaskawa, however, about competition in the United States from the Hitachi-General Electric combination that was announced in August 1981.

Other cooperative ventures include the Toistechnique Company of Sweden (1978), Messergresheim of West Germany (1979), and Alcas of Italy (1980).

Concomitant to Yaskawa's export orientation is its continuing emphasis on research and development. Yaskawa is one of twenty firms participating in the seven-year national flexible manufacturing system development being run under the aegis of MITI's Agency of Industrial Science and Technology. Yaskawa continues to emphasize technological agreements with companies having strong engineering capabilities (Feado Robot Systems in Japan, Barrons Engineering in South Africa, and A.N.I. Perkins of Australia).

Despite the expected success of Yaskawa's arc-welding robots, two problems remain. The shape of parts of robots used in the automotive parts industry tends to be very complicated, using a crooked welding line which reaches where a television camera cannot see. Secondly, cost is a problem. Currently, the minicomputers on the robot alone cost more than 10 million yen; a price which is too high to be marketable. If Yaskawa can solve these two problems, it will remain a leader in the robotics industry.

The company plans over the long run for a 2:1 ratio of domestic to export business, so the movement toward foreign technological agreements is likely to continue. In addition to foreign agreements, Yaskawa established in 1979 the Yaskawa Trading Corporation to concentrate on sales of robots, thereby expanding its distribution channels from distributors of welding supplies only.

EUROPE

ASEA

Overview

ASEA, a leading Swedish company in the heavy electrical parts industry, is currently a leader in the European AMES market. ASEA's robot, initially introduced in 1974, was the first all-electric and therefore clean and silent robot. ASEA's strength in the robotics industry continues to be its emphasis on providing a broad product range utilizing inhouse technology. It has grown through mergers (e.g., with Electrolux) and has made a strong commitment to the U.S. market.

Current Position

ASEA's robot division accounts for about 1.5 percent of its total turnover and its current production capacity is 900 robots per year. Sales are increasing 70 to 80 percent yearly and ASEA worldwide robot sales were $40 million in 1981. ASEA already has 2,000 robots operating worldwide since introducing its first robot in 1974. Domestic sales account for only 10 percent of ASEA's total robot sales; West Germany 25 percent, the United States 25 percent and the rest of the world 40 percent.

The ASEA robot is typically applied in spot welding, arcwelding, glueing, inspection, finishing, polishing, and material handling. ASEA's robotics division provides, in addition to robots, peripheral systems, process engineering help, service and replacement parts, on-site installation assistance, customer training, and financing support.

ASEA's strength as a major European robotics firm has been its ability to achieve manufacturing efficiencies which have allowed it to reduce the price of some of its robots by over 20%. In an attempt to gain a larger world market share, ASEA has emphasized development of a comprehensive program to meet customer requirements for flexible manufacturing solutions. New robot sizes and versions have been introduced, a major accessory program has been developed, and training resources have been expanded.

Corporate Strategy

ASEA's corporate strategy emphasizes the broadest possible product range, as the company believes this provides for competitiveness and is a prerequisite for growth. Research and development work is done in close cooperation with major customers and specialist companies.

ASEA accounts for about one tenth of the overall research and development activities of Swedish industry. In the past few years, the company has been granted more patents than any other Swedish firm. ASEA already has two robotic engineering centers and is adding two more.

Recently, ASEA merged with the Swedish Electrolux robotics division. This merger complements ASEA's range of products and expands its global marketing network.

ASEA is now completing a manufacturing plant in Milwaukee which will provide additional robots for the U.S. market.

A new training center was built at corporate headquarters in Vasteras, Sweden in 1982, exemplifying increasing concern about problems of costs and manpower in user support services. It was developed in order to enable users to fully utilize all the features provided by ASEA's robot technology. Training is also offered in West Germany, England, the United States and a number of other countries.

ASEA maintains a worldwide service network for robots to perform installation, commissioning and servicing assignments as well as doing plant project work.

OLIVETTI

Olivetti, one of the largest companies in Italy, has initiated a radical technological changeover from mechanical technology to electronic technology, including robotics. In doing so, Olivetti has applied two fundamental policies; heavy investment in research (which has tripled in four years, especially in electronics) and an emphasis on the creation of a dense network of agreements, acquisitions and joint ventures with leading companies in specific technological areas where Olivetti could not initiate its own research.

This strategy is designed to acquire first place positions in various specific product and technology areas, or follower positions in some areas, by protecting and acquiring high market shares with strong commercial organization barriers.

Olivetti recently signed a five-year contract with Westinghouse. It provides initially for the acquisition of 40 Olivetti Sigma robots by Westinghouse, which will have the exclusive license for producing, distributing, and servicing Sigma robots in North America.

Olivetti's European marketing strategy targets small to medium users and specializes in computer controlled machinery centers for small batch production. In Europe, Sigma is the most widely used robot for use in precision machining, electromechanics and electronics production.

Olivetti has another joint venture with Allen Bradley Company which covers reciprocal marketing, sales, and service between the United States and Europe. The contract also provides for complementarity in design and engineering. Allen Bradley will supply large, sophisticated, high-cost machinery while Olivetti covers smaller, simpler equipment. Finally, there will be production investment collaboration between the two companies.

RENAULT

Renault, France's second largest industrial company and the world's seventh largest auto maker, is also the leading French producer of robots. It is the only French company offering a complete series of robots including intelligent robots.

Renault has deployed 254 robots, largely in its automotive plant at Davai in northern France. The latest robots developed by Renault have enabled it to produce an auto in five hours less than previously possible. The latest robots can also be reprogrammed to vary production according to demand within the same day.

A shape-recognizing intelligent robot was developed by Renault's Division of Advanced Technologies and Automation and was installed in August 1981 at Renault's Cleon plant.

To install robots, Renault undertook a large training program which involved the expenditure of 400 million francs. While the introduction of robotics did not eliminate any jobs in Renault plants, it did raise qualifications and salaries. Renault's robots are utilized in welding, painting, spraying, cutting, and in transporting heavy parts.

Renault has also developed the first French flexible workshop at the Renault Industrial Vehicles plant where truck transmissions are made. This facility represents a considerable investment for Renault, 45 million francs, including 6 million francs in assistance from the Director of the Electronics and Data Processing Industries in the form of discounted loans. Only 15 people (three shifts of five) are required to ensure normal operation of the workshop. Operating costs are slightly less than formerly, although maintenance costs are slightly higher. Renault believes the newer facility will allow it to combine productivity with flexibility, reaching a productivity level close to that achieved by large batch manufacturing while producing in small and medium sized batches.

In October 1980 Renault formed a joint venture with Ransburg Corporation (United States) for the development, manufacturing, and distribution of industrial robots. Ransburg owns 51 percent of the venture and Renault 49 percent. Renault is attempting to increase its global market share in the robotics field and believes that the current agreement with Ransburg will establish a marketing base to expand from in the United States.

Renault's success in the robotics industry is due in part to the lack of government interference in management, including any pressure to change investment policies. Renault's philosophy has been to solve its own automation problems without waiting for the results of government research programs which the company believes have little connection with industrial reality. Renault's management also believes that progress in automation can only be achieved with the support of labor, and is acting to improve working conditions as well as reduce costs.

TRALLFA

Trallfa, the leading Norwegian producer of robots, has emphasized the development of flexible automation. With 15 years of experience and over a thousand successful installations of robots, Trallfa is working on applying computer control to the total production process. Trallfa's most recent robot, the TR-4000, offers increased program capacity, greater opportunities for mixed production with a high degree of utilization on the manufacturing line, along with lower costs for intermediate materials storage. The TR-4000 is chiefly applied in spray painting and welding. Trallfa can supply the TR-4000 as a complete "welding package" which can achieve economic efficiency even on small batches.

Trallfa robots are produced in several locations with licensed production taking place in Japan and the United States. Servicing is worldwide and Trallfa has established a European head office in Germany to provide better service for customers and distributors in Europe.

INDEX